INORGANIC MATRIX COMPOSITES

INORGANIC MATRIX COMPOSITES

Proceedings of the Discussion Meeting sponsored by
Jawarharlal Nehru Center for Advanced Scientific Research
and the Structural Materials Division of TMS
held at the Indian Institute of Science, Bangalore, India
March 8-11, 1995

Edited by

M. K. SURAPPA
Associate Professor
Department of Metallurgy
Indian Institute of Science
Bangalore - 12

A Publication of

Minerals • Metals • Materials

A Publication of The Minerals, Metals & Materials Society
420 Commonwealth Drive
Warrendale, Pennsylvania 15086
(412) 776-9000

The Minerals, Metals & Materials Society is not responsible for statements or opinions and is absolved of liability due to misuse of information contained in this publication.

Printed in the United States of America
Library of Congress Catalog Number 96-75344
ISBN Number 0-87339-323-6

Authorization to photocopy items for internal or personal use, or the internal or personal use of specific clients, is granted by The Minerals, Metals & Materials Society for users registered with the Copyright Clearance Center (CCC) Transactional Reporting Service, provided that the base fee of $3.00 per copy is paid directly to Copyright Clearance Center, 27 Congress Street, Salem, Massachusetts 01970. For those organizations that have been granted a photocopy license by Copyright Clearance Center, a separate system of payment has been arranged.

© 1996

If you are interested in purchasing a copy of this book, or if you would like to receive the latest TMS publications catalog, please telephone 1-800-759-4867.

CONTENTS

Foreword vii

MICROSTRUCTURE, RHEOLOGICAL BEHAVIOUR AND FORMING IN THE
SEMI-SOLID STATE OF SiC REINFORCED ALUMINUM MATRIX COMPOSITES 1
    M. Suery

SOLIDIFICATION STUDIES IN ALUMINUM MATRIX COMPOSITES 15
    B. Dutta, S. Mannikar and M. K. Surappa

FABRICATION AND MECHANICAL PROPERTIES OF IN SITU FORMED
CARBIDE PARTICULATE REINFORCED ALUMINUM COMPOSITE 31
    T. Choh

THE DEVELOPMENT OF CAST METAL MATRIX SILICON
CARBIDE PARTICLE REINFORCED COMPOSITES 41
    B. Inem

FEASIBILITY OF ULTRASONIC INFILTRATION IN PREPARATION
OF METAL MATRIX COMPOSITES 59
    J. Pan, D. M. Yang and H. Wan

CASTING OF COMPOSITE COMPONENTS 69
    S. Ray

METAL MATRIX COMPOSITES FABRICATED BY PRESSURE ASSISTED
INFILTRATION OF LOOSE CERAMIC POWDER 91
    M. A. Taha and N. A. El-Mahallawy

EFFECT OF TYPE OF PROCESSING ON THE MICROSTRUCTURAL
FEATURES AND MECHANICAL PROPERTIES OF A SiC REINFORCED
Al-Cu METAL MATRIX COMPOSITE 109
    M. Gupta, L. Lu, M. O. Lai and A. S. Ee

HIGH TEMPERATURE DEFORMATION OF MAGNESIUM AND ALUMINUM
ALLOY COMPOSITES REINFORCED BY SiC WHISKERS 121
    M. Sugamata

TECHNOLOGY DEVELOPMENT OF CAST
ALUMINUM BASED COMPOSITES 131
    B. C. Pai, R. M. Pillai and K. G. Satyanarayana

STRUCTURE-PROPERTY RELATIONSHIPS FOR LAYERED DRA MATERIALS 155
    T. M. Osman, J. J. Lewandowski, W. H. Hunt and D. R. Lesuer

PROCESSING OF A PARTICULATE METAL MATRIX
COMPOSITES BY ROLL BONDING METHOD 165
    Y. P. Yao and W. B. Lee

TRIBOLOGICAL STUDIES ON ALUMINUM MATRIX COMPOSITES.................... 171
    B. N. Pramila Bai, R. A. Saravanan and M. K. Surappa

INFILTRATION OF Al_2O_3-Al COMPOSITES INTO
COATED SILICON CARBIDE 193
    V. Jayaram, S. Kumar, T. V. Mani, M. S. M. Saifullah,
    J. Sarkar and K. G. K. Warrier

COMBUSTION SYNTHESIS OF $MoSi_2$ BASED COMPOSITES 207
    J. Subrahmanyam

DISCONTINUOUSLY REINFORCED TITANIUM MATRIX COMPOSITES
VIA COMBUSTION ASSISTED SYNTHESIS 227
    S. Ranganath

PRODUCTION AND CHARACTERIZATION OF MMC TUBES 245
    A. K. Gupta, R. Sikand, R. C. Anandani, A. Dhar and I. A. Malik

DISCONTINUOUSLY REINFORCED PM PROCESSED MMCs
ACTIVITY AT THE DEFENCE METALLURGICAL RESEARCH
LABORATORY, HYDERABAD - 500 058, INDIA 269
    B. V. R. Bhat, V. V. Bhanuprasad, R. Mitra, M. K. Jain,
    A. B. Pandey and Y. R. Mahajan

AUTHOR INDEX 305

FOREWORD

Inorganic Matrix Composites comprising both Metal and Ceramic Matrix Composites have reached an advanced stage of development in the laboratory. However, their commercial exploitation has been slow. They have begun to appear in selected items in the automotive and aerospace industries. We still continue to have many fundamental research issues of properties/structure as well as problems relating to processing. The main objective of the Discussion Meeting on "Inorganic Matrix Composites" held at the Indian Institute of Science (IISc), March 9-11, 1995 in Bangalore, was to bring together researchers in different disciplines of Inorganic Matrix Composites to highlight and exchange the results of their studies. The in-depth discussions and technical interface during the meeting, reflected the wide range of interest in this interdisciplinary field.

The venue of the Discussion Meeting has an added significance since the early work on cast MMCs was nucleated at the Indian Institute of Science more than two decades ago. The process of synthesizing aluminum based MMCs by adding pretreated ceramic particles was developed at IISc by Professor P. K. Rohatgi and his co-workers.

The 18 papers contained in this volume present the state of the art in the areas of processing, structure/property relations and industrial applications of Inorganic Matrix Composites.

The discussion meeting was sponsored by the Jawaharlal Nehru Center for Advanced Scientific Research (JNCASR). I express my sincere thanks to Professor C. N. R. Rao for his support. This meeting was cosponsored by TMS and I thank Dr. Om Arora of SMD for this. I would also like to thank all the authors for the timely submission of papers which has facilitated the prompt publication of the proceedings. Finally, I thank Ms. Wendy McCalip and other staff at TMS for handling the publication.

M. K. Surappa
January 1996

MICROSTRUCTURE, RHEOLOGICAL BEHAVIOUR AND FORMING IN THE SEMI-SOLID STATE OF SiC REINFORCED ALUMINIUM MATRIX COMPOSITES

Michel SUERY

INSTITUT NATIONAL POLYTECHNIQUE DE GRENOBLE
GENIE PHYSIQUE ET MECANIQUE DES MATERIAUX
Unité de Recherche Associée au CNRS n°793
ENSPG, BP 46, 38402 SAINT MARTIN D'HERES CEDEX, FRANCE

ABSTRACT

A356 Al alloys reinforced with 20 vol% of SiC particles showed globularisation of the microstructure when maintained in the semi-solid state and subsequent coarsening of the globules. This globularisation is associated with a decrease of the stress during compression which becomes of the order of 0.1 MPa for 30 to 40 min. holding. Such low stresses allowed the composites to be forged isothermally in the semi-solid state to produce specimens for mechanical property evaluation. Forging led then to fully homogeneous specimens without any segregation of the liquid phase. This results to good mechanical properties, even better than those of as-cast composites. The experiments thus demonstrated that forming of composites can avantageously be carried out in the semi-solid state owing to the very low forming efforts that are required when the structure of the solid phase has become globular.

INTRODUCTION

Forming in the semi-solid state is now used quite extensively for aluminium alloys owing to its advantages over conventional casting and forging such as reduction of macrosegregation and porosity, achievement of near net shape parts and low forming efforts (1-6). These advantages require a globular non dendritic structure of the solid phase which can be obtained generally by stirring during prior solidification of the alloy. This structure which is very different from the ordinary dendritic one explains the particular behaviour of the material when subjected to shear deformation during the forming operation.

It has been demonstrated recently that this kind of microstructure can also be generated in the presence of ceramic particles (7-12). In addition to the previously mentioned advantages, forming of aluminium matrix composites in the semi-solid state may then overcome problems like rapid tool wear, breakage and segregation of the reinforcing elements and reduce machining of difficult to machine materials.

The aim of this paper is to report some results concerning microstructure, rheological behaviour and forming in the semi-solid state of A356 aluminium alloys containing 20 vol.% SiC particles. Compression tests were particularly carried out at various strain rates to determine the behaviour of the composites as a function of the holding time in the semi-solid state. The forming tests were performed isothermally to produce specimens for mechanical property evaluation.

MICROSTRUCTURE OF THE COMPOSITES IN THE SEMI-SOLID STATE

The material used for this investigation was provided by DURALCAN, San Diego, USA, in the form of ingots and consists of an A356 alloy reinforced with 20 vol% SiC particles of 13 µm average diameter (Material F3A20S). The microstructure of the as-received material is shown in Figure 1. Uniform distribution of the particles is obtained at the scale of the specimen. However, at larger magnification, segregation is observed because the particles have been pushed out by the solidification front; they are thus preferentially located in the eutectic mixture. This segregation is particularly important owing to the slow cooling rate of the ingot which led to relatively coarse dendrites of the primary Al-rich phase. Moreover some porosities are often observed in the microstructure so that the ingot was fully remelted at about 700°C and poured in a preheated (300°C) steel mould in which solidification was accomplished under a pressure of 100 MPa. This procedure leads then to sound ingots without any porosity. In addition, the microstructure has been refined compared to the initial one (figure 2). Segregation of particles in the eutectic mixture is still observed but the dendritic structure of the primary phase has been strongly affected by the presence of the particles. All the experiments were carried out with this material.

Holding the composites in the semi-solid state leads to globularisation of the microstructure. Figure 3 shows the microstructure of the composite after holding for 40 min. at 585°C. A quasi-globular structure of the primary Al rich phase has been formed with SiC particles concentrated in the eutectic mixture, but some globules are still agglomerated. Increasing the holding time in the semi-solid state leads to the coarsening of the globules as shown in figure 4 which represents the variation with holding time of the average globule diameter measured by the linear intercept method on a Log-Log scale. The slope of the straight line obtained in this plot is close to 0.17 which indicates that coarsening of the globules is strongly affected by the presence of the particles. Indeed the slope of the curve for an unreinforced alloy is usually larger close to 0.33. The presence of the particles homogeneously distributed in the liquid seems thus to hinder coarsening of the globules. The two mechanisms of coarsening are probably affected: coalescence and Ostwald ripening. Coalescence is reduced owing to the fact that the globules are less agglomerated due to the presence of the particles. Ostwald ripening is altered because of the reduction of the diffusion coefficient in the liquid.

RHEOLOGICAL BEHAVIOUR

The rheological behaviour of the composites in the semi-solid state was investigated by compression tests between two parallel plates. The tests were carried out at constant strain rate at 585°C. Figure 5 shows the influence of the holding time at 585°C before compression on the compression stress of the composites. For these experiments, the strain rate was $0.1\ s^{-1}$ and the holding time varied from 2 to 40 min. The figure shows that the compression stress decreases with increasing holding time becoming of the order of 0.1 MPa for 40 min. holding. This decrease is associated with the globularisation of the microstructure which is nearly completed after 30 or 40 min.. For this time, deformation of the specimen is fully homogeneous without appreciable deformation of the individual globules as demonstrated by comparison of the microstructure of the deformed specimen (figure 6) with that of the undeformed one for the same holding time. These results are interesting for forming applications since they show that holding of a composite in the semi-solid state for about 30 min. is sufficient to reduce considerably the deformation stresses. Such a reduction is even more important for composites extruded before partial remelting. Indeed in this case the compression stresses can be as low as a few kPa for the same strain rate (11). Moreover, holding time for only 5 min. is sufficient to reach such low deformation efforts.

In order to investigate the influence of the strain rate on the rheological behaviour of the material, compression tests were carried out at various strain rates after a holding time of the specimens before compression of 40 min.. Figure 7 shows the variation with strain rate of the compression stress on a Log-Log scale. A linear relationship is obtained, associated with a constant value of the strain rate sensitivity parameter m over the strain rate range investigated.

This value of m is close to 0.45 characteristic of a pseudo-plastic behaviour. A similar behaviour was observed with composites extruded before partial remelting (Figure 7)(11) together with composites based on a Al-Cu-Mg alloy in both the as-cast and extruded conditions (8).

FORMING IN THE SEMI-SOLID STATE

The forming experiments were carried out isothermally to produce specimens for mechanical property evaluation. Two types of die were used:
- a graphite die inserted into a steel liner which produces a single specimen (15 mm diameter, 70 mm length) each time
- a stainless steel die, based on the same principle as the previous one which is able to produce 4 specimens (14 mm diameter, 65 mm length) during each run.

The die was heated isothermally at the required temperature using a 3 kW electrical resistance. The billet was put into the container of the die with an insulating sheet on top of it to achieve and maintain a constant temperature. After a given holding time in the semi-solid state, the insulator was replaced by a graphite piston heated separately at the same temperature as the billet. The punch of the press was then activated at about 25 $mm.s^{-1}$ to force the semi-solid metal into the cavities of the die. Small holes drilled in the bottom of the cavities allowed air to escape in order to produce sound specimens. The applied load was such that it corresponds to a pressure of about 35 MPa. The forged part was thereafter removed from the die. Figure 8 shows a typical part produced with the stainless steel die and Figure 9 shows the microstructure of the composite at three different locations along the produced specimens. For this figure the holding time before forging was about 20 min for which the microstructure is not yet globularised. Nevertheless, homogeneous deformation was achieved. Increasing the holding time before forging leads to a more globular microstructure as shown in figure 10 which corresponds to a specimen produced in the graphite die after a holding time of 60 min. No segregation is observed along the specimen although some evidence of deformation of the globules appears in figure 10(a). This means that no preferential liquid flow has occurred during the forging process. Increasing the holding time up to 120 min. was found not to affect the formability of the composites, although some globule coarsening has occurred. However the dimension of the specimen is still much larger than the size of the globules which explains the absence of liquid segregation during forging.

MECHANICAL PROPERTIES AFTER FORMING

Tensile specimens were machined from the parts produced by isothermal forging at 585°C using the graphite die. Four series of specimens were investigated corresponding to four different holding times in the semi-solid state before forging, namely 30, 60, 90 and 120 min. The specimens were tested either in the as-forged condition or after heat treatment (12 h. solutionizing at 540°C, water quenching followed by aging for various times at 160°C). In the as-forged condition, however, large scattering was observed in the results owing to the fact that cooling of the specimens after forging was not controlled. For comparison, tensile specimens machined from as-cast (solidified under a 100 MPa pressure) or extruded (16:1 ratio at 450°C) Duralcan composites were also tested. All the tests were performed at a strain rate of 10^{-3} s^{-1}. The results are as follows:

- The holding time of the billet at 585°C before forging does not significantly affect the stress-strain curve of the forged specimens (figures 11 and 12);
- The forged composites show slightly larger fracture stress and fracture strain than the as-cast composites (figure 13). However, these elongations remain much smaller than those observed with the extruded materials. Also the extruded composites exhibit higher fracture stress (figure 12). The properties of the composites forged in the semi-solid state are thus intermediate between those of the as-cast and the extruded composites.
- The materials are sensitive to an aging treatment at 160°C (figure 14). The maximum stress increases initially with increasing aging time after which it levels off or slightly decreases. Whatever the treatment, the forged composites show intermediate properties between the as-cast and the extruded composites.

These results thus show that isothermal forging in the semi-solid state leads to sound specimens with good mechanical properties, even better than the as-cast materials.

CONCLUSIONS

Holding an Al-Si based composite in the semi-solid state leads to globularisation of the primary solid phase with particles concentrated in the eutectic mixture. When the structure is globularised, the composites deform homogeneously under low stresses and exhibit a pseudo-plastic behaviour. Forming in the semi-solid state has been successfully achieved to produce specimens for mechanical property evaluation. These properties are better than those of cast composites which thus demonstrates the potential advantages of semi-solid forming for the fabrication of sound parts.

ACKNOWLEDGEMENTS

The author would like to thank Dr. L. Nguyen Thanh for having performed most of the experiments of forging and F. Pelloux for his technical help. This work was financially supported by the Commission of the European Communities, Brussels, under Brite/Euram contract n° BREU-0151C.

REFERENCES

1. M. KIUCHI, S. SUGIYAMA and K. ARAI: Proc. of the 20th International Machine Tool Design and Research Conference, 1979, pp. 71-78

2. M. KIUCHI, D. SUGIYAMA and K. ARAI: Proc. of the 20th International Machine Tool Design and Research Conference, 1979, pp. 79-86

3. M.C. FLEMINGS: Metall. Trans., 1991, vol. 22B, pp. 269-293.

4. K.P. YOUNG: Nature and Properties of Semi-Solid Materials, Ed. by J.A. Sekhar and J.A. Dantzig, TMS, Warrendale, 1991, pp. 245-266.

5. M.P. KENNEY, J.A. COURTOIS, R.D. EVANS, G.M. FARRIOR, C.P. KYONKA, A.A. KOCH and K.P. YOUNG, Metals Handbook, 9th Edition, ASM International, Metals Park, OH, 1988, vol. 15, pp. 327-338.

6. L.N. ZHANG, S.Q. WANG, M.F. ZHU, N. WANG and S.D. WANG: J. Mater. Proc. Technol., 1994, vol. 44, pp. 91-98.

7. M.A. BAYOUMI and M. SUERY: Proc. Int. Symp. "Cast Reinforced Metal Composites" Ed. S.G. Fishman and A.K. Dhingra, ASM International, Metals Park, OH, 1988, pp. 167-172.

8. L. NGUYEN THANH and M. SUERY: Proc. 2nd Int. Conf. on "Processing of Semi-Sold Alloys and Composites", Ed. S.B. Brown and M.C. Flemings, MIT Press, Cambridge, MA, 1992, pp. 427-435.

9. W.A. MAY, S.P. MIDSON and K.P. YOUNG: Ibid. pp. 390-397.

10. T.G. NGUYEN, M. SUERY and D. FAVIER: Mater. Sci. Eng., 1994, vol. A183, pp. 157-167.

11. L. NGUYEN THANH and M. SUERY: Mater. Sci. Technol., 1994, vol. 10, pp. 894-901.

12. W.H. KOOL, C.J. QUAAK and M. SUERY: Trans. A.F.S., in press.

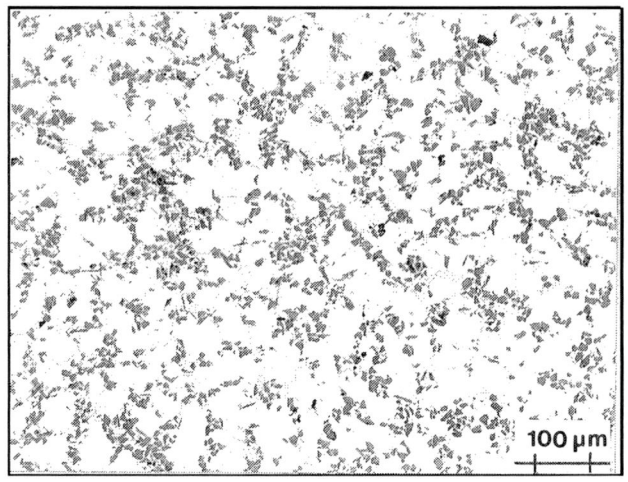

Figure 1: Microstructure of the as-received Duralcan composite.

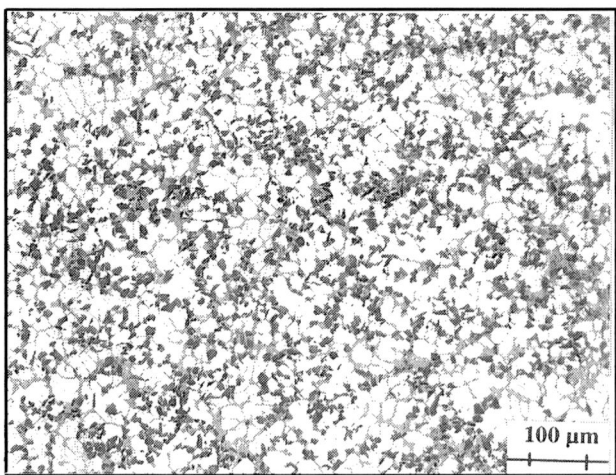

Figure 2: Microstructure of the Duralcan composite in the as-cast condition solidified under a 100 MPa pressure in a preheated (300°C) steel mould.

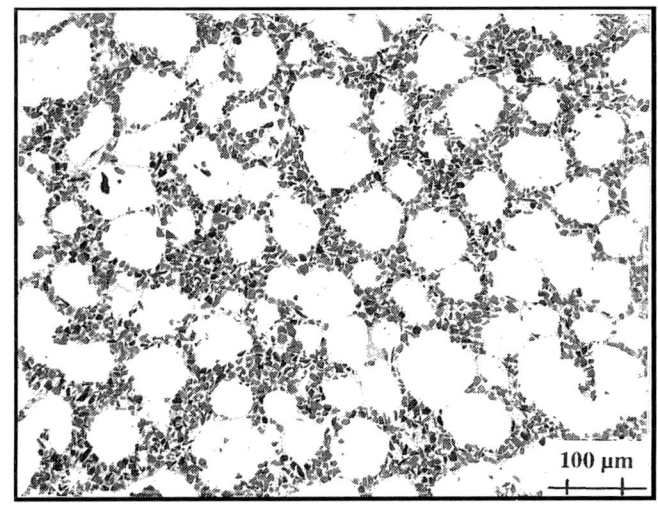

Figure 3: Microstructure of the composite after an isothermal holding of 40 min. at 585°C.

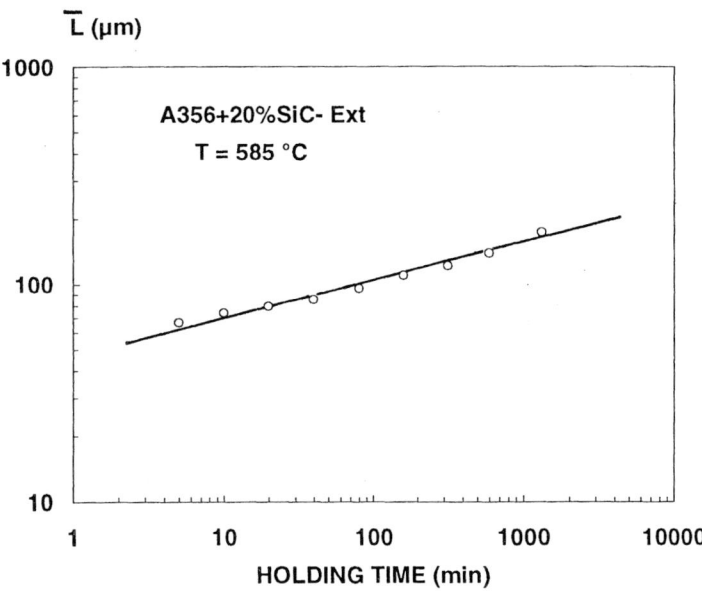

Figure 4: Variation of globule size $\overline{L}$ with holding time at 585°C for the Duralcan composite.

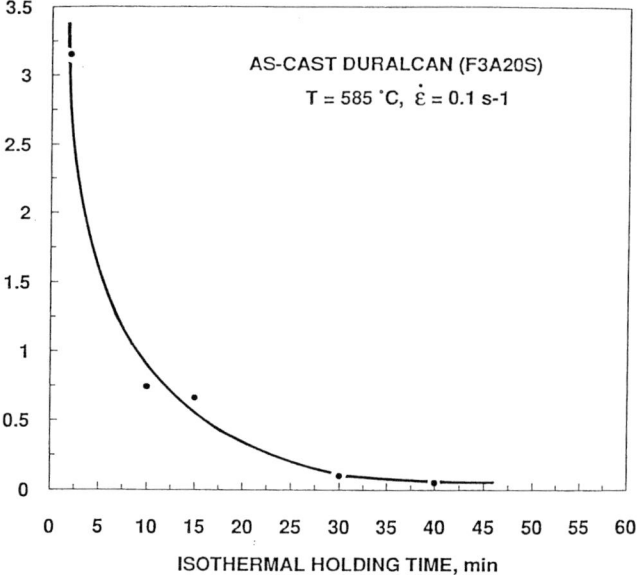

Figure 5: Compression stress versus holding time at 585°C before deformation.

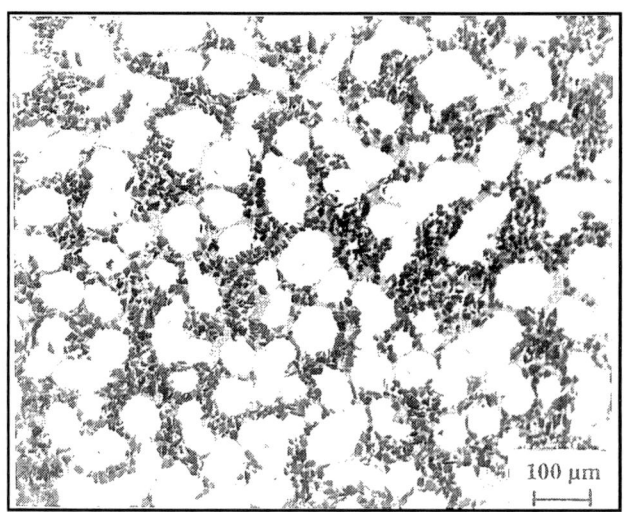

Figure 6: Microstructure of a composite specimen deformed in compression at 10^{-1} s^{-1} after holding at 585°C for 40 min..

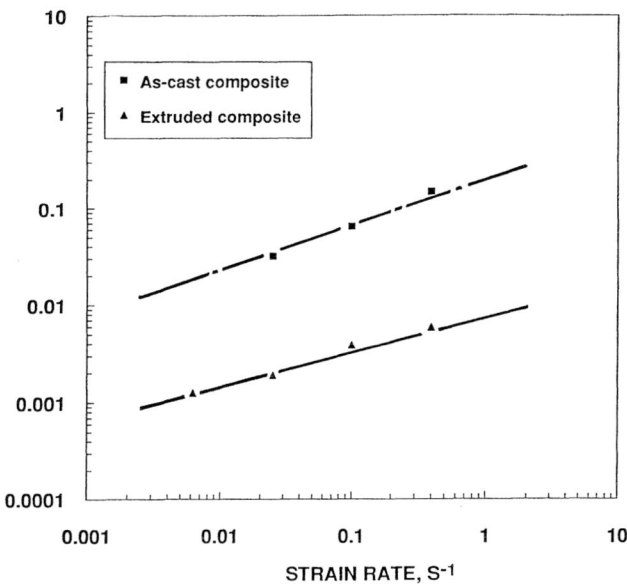

Figure 7: Strain rate dependence of the compression stress for composite specimens held for 40 min. at 585°C before deformation.

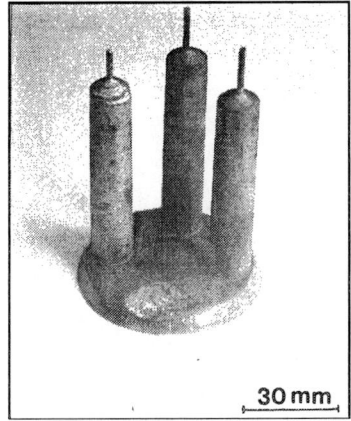

Figure 8: Typical forged part produced with the stainless steel die. A specimen has been cut

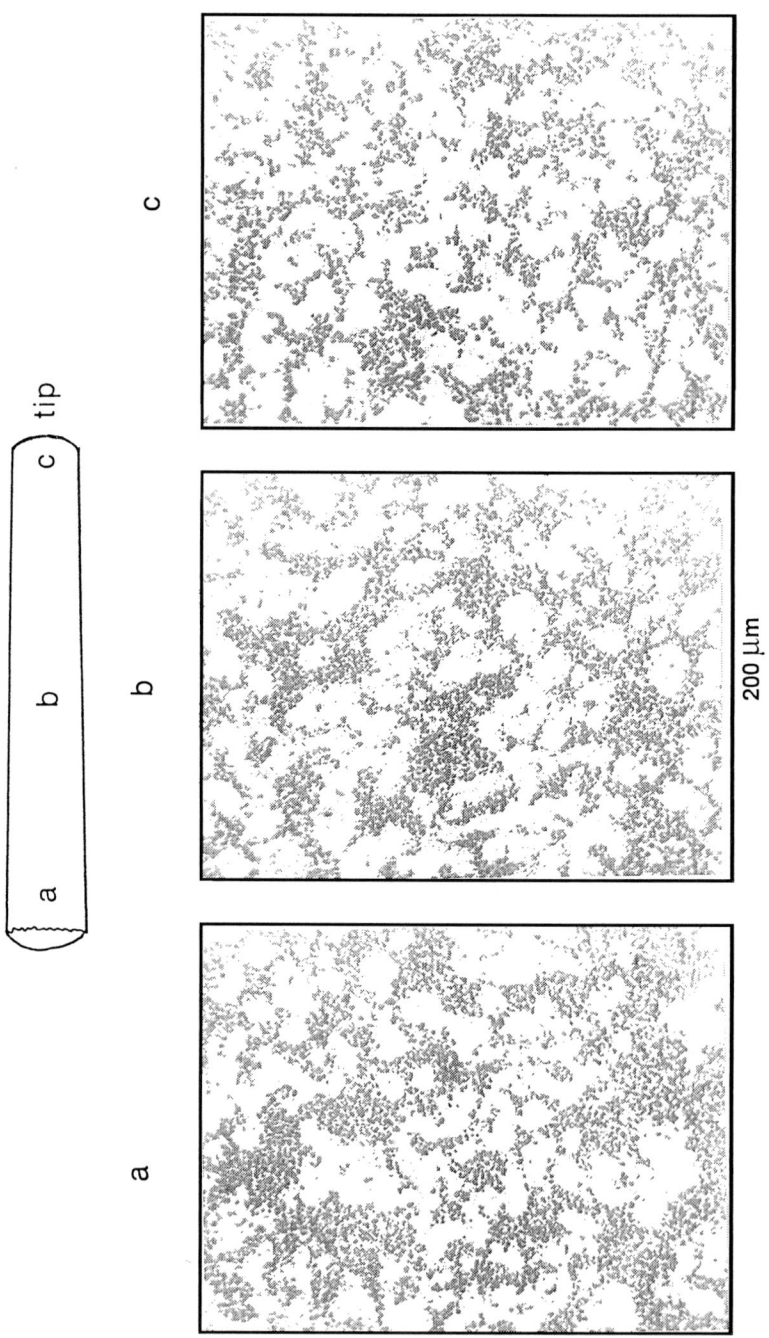

Figure 9: Microstructures of a forged specimen in the stainless steel die showing uniform distribution of the SiC particles. The holding time at 585°C before forging is 20 min..

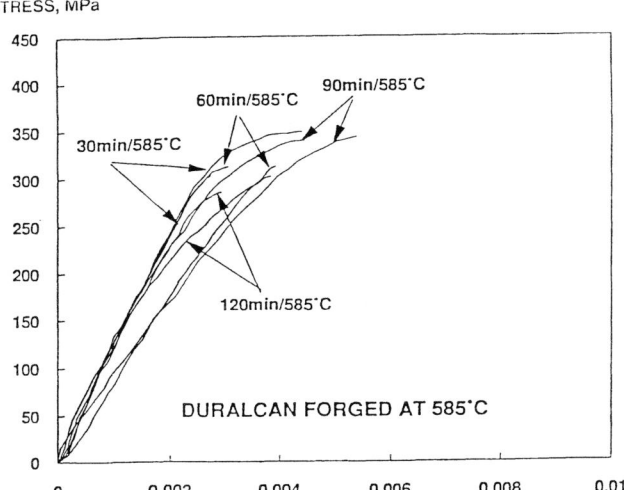

Figure 11: Stress-strain curves for composite specimens forged in the graphite die after various holding times at 585°C. After forging the specimens were solutionised for 12 h. at 540°C, quenched in water and aged for 24 h. at 160°C.

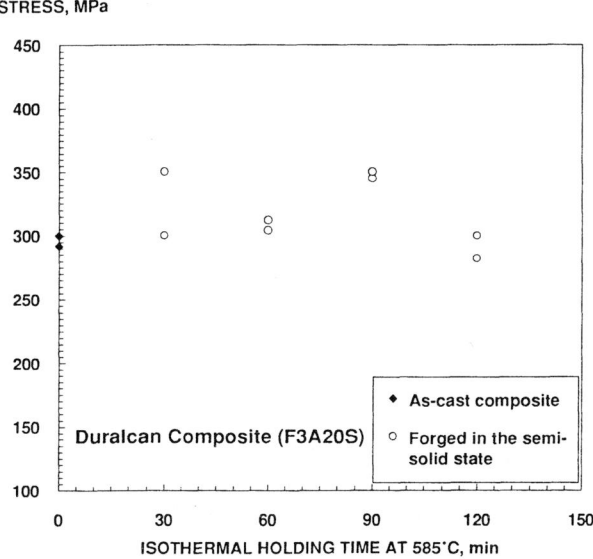

Figure 12: Influence of the isothermal holding time before forging on the maximum tensile stress of the forged specimens. After forging the specimens were solutionised for 12 h. at 540°C, quenched in water and aged for 24 h. at 160°C.

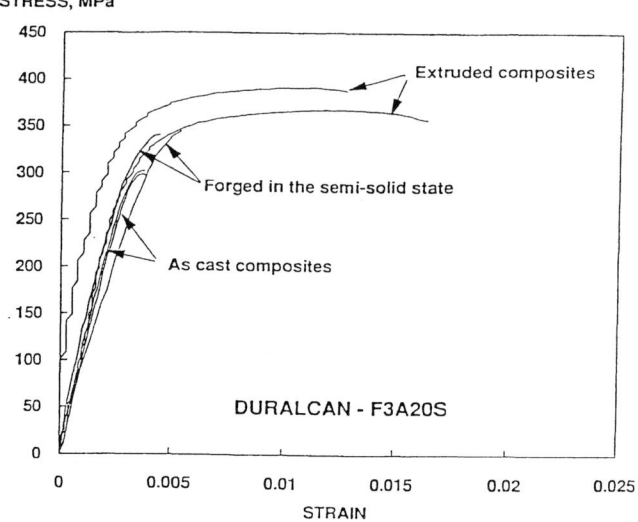

Figure 13: Influence of the forming process on the stress-strain curves for composite specimens. After forging the specimens were solutionised for 12 h. at 540°C, quenched in water and aged for 24 h. at 160°C.

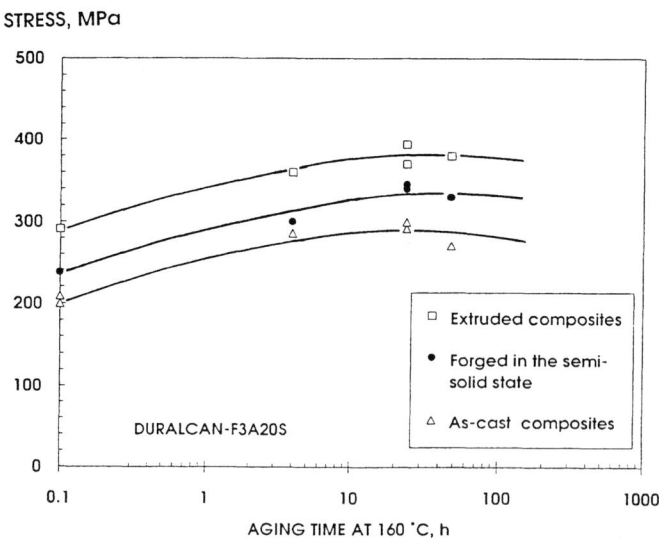

Figure 14: Influence of the aging time at 160°C on the maximum tensile stress for composite specimens produced by various forming processes.

SOLIDIFICATION STUDIES IN ALUMINIUM MATRIX COMPOSITES

B. Dutta, Sucheta Mannikar and M. K. Surappa

Department of Metallurgy
Center for Advanced Study
Indian Institute of Science
Bangalore 560 012

ABSTRACT

There have been increased efforts both on the experimental and theoretical front to understand the solidification phenomena occurring during processing of aluminium matrix composites. Much of these studies have been directed to gain knowledge about the conditions under which ceramic dispersoids are incorporated into different microstructural constituents. Further these studies have generated a data base on nucleating potency of reinforcement and accompanying changes in the grain structure of solidified composites. This paper reports the results of studies on solidification of aluminium alloy melt containing SiC particles relating to : a) morphology of advancing solidification front b) distribution of particles in the solidified matrix phase and c) particle interaction with the solute diffusion field in the vicinity of an advancing solidification front.

INTRODUCTION

There have been numerous studies aimed at understanding the microstructure evolution during solidification processing of MMCs. Processing conditions under which discontinuous ceramic reinforcements are incorporated into the solidifying metal/alloy matrix are still not clearly understood. The particle and the advancing solidification front may interact in different ways resulting in different microstructures. A number of experiments in this field have been carried out on transparent organic systems (where in-situ observation is possible) with inert particles. However, in case of metallic systems, the particle-solidification front interactions have to be inferred on the basis of microstructural observations i.e. the particle distribution as observed in the final microstructure.

The resulting effect of particle/solid-liquid interface interaction could be of three types ; engulfment, entrapment or pushing. Particle pushing over large distances by the growing interface into the last solidifying regions of the casting leads to macrosegregation. This feature is observed in castings solidified unidirectionally having either planar or cellular or dendritic interface. A particle is said to be engulfed when it is surrounded completely by the primary matrix phase. This results in a uniform particle distribution in case of all morphologies of the growth front and this type of distribution is generally most desired. For planar front, only pushing or engulfment is possible. However, in case of the cellular and dendritic fronts, there are additional possibilities. Particles may be pushed to the cell/dendrite boundaries and remain mechanically locked in those positions i.e. entrapment of particles can occur. This kind of segregation in intercellular/interdendritic regions is known as microsegregation. In case of dendritic solidification, the secondary dendrites also interact with the particles. Particle pushing by the secondary dendrites also leads to microsegregation of particles.

Theoretically, many attempts have been made to predict the criteria for the transition from pushing to engulfment. This could, in principle, facilitate the fabrication of composites having desired distribution of reinforcement by choosing appropriate solidification conditions. Most of these are based on critical velocity criteria evolved by balancing the attractive and repulsive forces acting on the particle. If the growth velocity is higher than this critical velocity, particles will be engulfed while a lower growth velocity will lead to particle pushing by the solidification front. Magnitude of these forces are calculated using different physico-chemical properties of materials. The model proposed by Uhlmann, Chalmers and Jackson [1] predicts the critical velocity as

$$V_{cr} = \frac{(n+1)}{2}\left(\frac{La_0\Omega D_l}{kTR^2}\right) \qquad (1)$$

The model is based on mass diffusion in the liquid gap between the particle and the solid which arises from variations in interface curvature. This model, however does not take into account interfacial energy and viscous drag on the particle. Also, a spherical particle shape has been assumed.

The Bolling and Cisse model [2], gives critical velocity value as :

$$V_{cr} = \left(\frac{4\psi(\alpha)kTa_0\sigma_{sl}}{9\pi\eta^2 R^3}\right)^{1/2} ; R > R_b \qquad (2)$$

taking interface shape as well as surface roughness of particles into consideration. However, in the model, thermal conductivities of the particle and the melt are assumed to be the same.

Postchke and Rogge [3] express critical velocity as a function of thermal conductivity difference and solute composition.

$$V_{cr} = \frac{1.3\Delta\sigma_0}{\eta}[16\left(\frac{R}{a_0}\right)^2\left(\frac{K_p}{K_l}\right)\left(15\frac{K_p}{K_l} + x\right) + x^2]^{-0.5} \qquad (3)$$

$$where, x = \frac{C_\infty \mid m_l \mid \Delta\sigma_0}{K_cG\eta D_l} \qquad (4)$$

Sasikumar and Ramamohan's model [4] suggests

$$V_{cr} = \frac{B}{6\pi\eta R^2}[0.127C_2^{0.4}C_3^{0.01} + 0.11C_2^{-0.04}C_3^{0.37}C_4^{0.64}] \qquad (5)$$

$$where, C_2 = \frac{B}{R^2\alpha} \qquad (6)$$

$$C_3 = \frac{\Delta SGR^4}{\Omega B} \qquad (7)$$

$$C_4 = \frac{1 - K_p/K_l}{2 + K_p/K_l} \tag{8}$$

This model takes into account the surface forces (due to which the repulsive force arises), the distortion of the temperature and solute concentration fields. However, this model too is valid only in case of non-dendritic interface.

Shanguan et al [5] have proposed that

$$V_{cr} = \frac{a_0 \Delta \sigma_0}{3\eta\alpha(n-1)R}\left(\frac{n-1}{n}\right)^n \tag{9}$$

This model accounts for the interfacial energy and the thermal conductivity.

Most of the models discussed above are applicable only for pure metals. In practice, in most of the cases, the matrix is an alloy. Sasikumar et al [4] have shown that the breakdown of plane front occurs before critical velocity is reached during solidification of AMCs. Hence all these critical velocity based models are not valid for MMCs based on alloy matrices.

In addition to the models discussed before, there are models for particle pushing based on changes in thermal fields in front of the advancing interface. Zubko's model [6] predicts engulfment/pushing based on thermal conductivity ratio (α) of the particle and the melt.

$$\alpha = \frac{K_p}{K_l} \tag{10}$$

If α is less than 1, particle pushing is predicted and if α is greater than 1, engulfment is expected to occur.

This was further modified by Surappa and Rohatgi [7] by taking heat diffusivity instead of thermal conductivity. The basis of prediction is the following ratio β.

$$\beta = \left(\frac{K_p C_p \rho_p}{K_l C_l \rho_l}\right)^{1/2} \tag{11}$$

Pushing to engulfment transition is predicted at $\beta > 1$.

Predictions based on these models do not match with the results available under wide ranging experimental conditions. These models assume plane front solidification conditions which may not be the case in reality. The applicability of some of the models under various experimental solidification conditions has been tested and reported in this paper. The macro and micro distribution of particulate under a range of solidification conditions has been qualitatively analyzed.

Further, the presence of particles leads to changes in solidification structure of the matrix. Sekhar and Trivedi [8] in their experiments with an organic binary system with inert particles have reported change in morphology of the binary matrix in presence of particles. Also, a change in the solute distribution is expected. This has not been reported in case of AMCs based on alloy matrices. In this paper we report results of our experiments carried out to study the changes in the solidification structure, solute diffusion and other associated effects due to presence of SiC particles during solidification of AMCs.

MATERIALS AND EXPERIMENTS

All the composites were prepared by the melt-stir casting technique. Details of matrix materials used, SiC particulate size and volume fraction and the solidification conditions employed in the experiments are listed in the Table 1.

Materials were solidified under two different conditions. Unidirectional solidification was carried out in refractory molds with copper chills at the bottom and exothermic powder at the top. In these cases G and V were not constant. In the second method, composites were solidified in a permanent mold (cast iron mold) where temperature gradient and cooling rate were measured by thermocouples placed along the central axis at different distances from the mold bottom. The interface velocity was calculated from the cooling curve.

The castings were cut and polished along the meridian planes for macro and microstructure observations. The Neophot optical microscope and JEOL SEM was used for this purpose. Cambridge SEM with EDAX was used to obtain X-ray elemental mappings.

RESULTS

1. CRITICAL VELOCITY PREDICTIONS AND PARTICLE DISTRIBUTION

Tables 2a and 2b show the results of the unidirectional solidification experiments carried out in insulating refractory molds. These results essentially describe the interaction between a growing secondary dendrite arm and the particle (Fig. 1). In this case local interface velocity (i.e growth rate of secondary dendrite arms) measurements were made by measuring the secondary dendrite arm spacing. The following relations were used :

$$DAS = 7.5 t_f^{0.39} \quad for \quad Al - Cu \tag{12}$$

$$DAS = 20.3 t_f^{0.33} \quad for \quad Al - Cu - Mg \tag{13}$$

$$V_i = \frac{DAS}{2 t_f} \tag{14}$$

This local interface velocity was calculated in the region where particles have been engulfed. These velocities and the computed critical velocity based on different models is shown in Table 2. Predictions of all but Shanguan's model seem to be in agreement with experimental observations. Table 3 shows the results of models based on thermophysical properties. Both the models predict pushing whereas engulfment has actually been observed.

In case of the permanent mold casting, the comparison between actual interface velocity (i.e growth rate of primary dendrites) and predicted critical velocities are shown in Fig. 2. The actual interface velocity was measured along the central axis of the casting. All the models, except Shanguan's model, predict critical velocities much lower than the actual interface velocity expected in the experiments i.e. entrapment is expected to occur in accordance with these models. However, Shanguan's model predicts values nearabout the experimental results.

Macrostructure of the castings is shown in Fig. 3. Pushing of particles to the last solidifying regions near the top can be seen in the directionally solidified composites prepared in refractory molds. In case of Al-Cu-SiC$_p$ composites cast in permanent mold, particle segregation in the form of an inverted "V" type particle depleted zone was observed as shown in Fig. 3b.

2. CHANGES IN MORPHOLOGY, GRAIN SIZE AND SOLUTE DIFFUSION FIELD

Presence of particles is found to influence the morphology of the growing interface. In case of the unidirectionally solidified composites, the particle free zone consisted of long columnar dendrites. However, in the top region of the castings where particles are present, the structure was predominantly cellular-dendritic or equiaxed-dendritic. This effect is shown in Fig.4. Similar effect was observed in the permanent mold composite casting (Fig.5).

In case of the permanent mold composite castings, both grain size and secondary DAS changes in presence of particles. Grain size as well as secondary DAS become coarser in presence of SiC$_p$ as compared to those in the unreinforced alloy cast under similar conditions. The variation in grain size both as a function of SiCp content and the distance from the mold bottom is shown in Fig. 6a. Variation of secondary DAS with distance from the mold bottom and SiC particle content is shown in Fig. 6b.

X-ray elemental mapping of Cu in the directionally solidified composite shows a thin layer or band of Cu in the form of semi-annular ring around the SiC particles. In the SEM micrograph (Fig. 7) the direction of solidification is from the right to the left. The semi-annular ring of about 4μm thick Cu layer, is seen only on the bottom side of the particle indicating instantaneous entrapment of solute rich layer by the growing solid.

In case of A356-SiC$_p$, heterogeneous nucleation of eutectic Si was observed as shown in Fig.8. However, nucleation of eutectic Si was not seen around every SiC particle. In addition sites for nucleation seem to be very limited. In case of Al-Cu and Al-Cu-Mg composites, no change in morphology of the surrounding eutectic was observed.

DISCUSSION

a) PARTICLE PUSHING OR ENTRAPMENT

The result of particle-solidification front interactions depends upon the balance of forces acting on a particle. The forces acting on the particle are the repulsive force, viscous drag and the buoyancy force.

The repulsive force arises due to the difference in surface energies between the particle and the matrix. If $\sigma_{ps} - (\sigma_{pl} + \sigma_{sl}) > 0$, the advancing interface exerts a repulsive force [3] on the particle as given by :

$$F_r = 128\pi B \frac{R_I^3 R^3 (R_I + R + d)}{d^2 (d + 2R_I)^2 (d + 2R)^2 (d + 2R_I + 2R)^2} \qquad (15)$$

If the particle is to be pushed ahead of the moving interface, liquid metal should be continuously fed into the gap between the particle and the moving solidification front. This will exert a viscous drag force [5] on the particle.

$$F_d = 6\pi\eta V \frac{1}{d^2}\left(\frac{R_I R}{R_I + R}\right)^2 \qquad (16)$$

The buoyancy force [5] arises on account of the density difference between the particle and the melt.

$$F_b = \frac{4}{3}\pi R^3 g(\rho_p - \rho_l) \qquad (17)$$

The net force acting on the particle is given by :

$$F_N = F_r - F_d - F_b \qquad (18)$$

This net force decides whether a particle is pushed or engulfed. Examining the equations 1, 2 and 3, it becomes clear that F_r is a function of interface radius (R_I) and gap thickness (d), while F_d is a function of R_I, d and V, other system parameters remaining constant. In case of plane front solidification, R_I does not change significantly with changing velocity, hence the F_r remains constant. However, F_d increases sharply with increasing velocity (Fig. 9a) and beyond a critical velocity, particle is engulfed by the growing solid. However, in case of dendritic solidification, interface radius (which is equivalent to dendrite tip radius here) decreases sharply with increasing velocity and hence F_r also decreases. F_d initially increases with increasing velocity and then starts decreasing as the effect of dendrite tip radius becomes predominant (Fig. 9b). This marks significant differences in the behaviour of the forces between plane front and dendritic solidification.

It is clear from the above discussions that the effect of dendrite tip radius needs to be considered while deriving the expression for pushing/engulfment transition velocity during dendritic solidification. Hence it is obvious that the plane front models do not give correct results when applied to dendritic solidification. The failure of the existing models to predict pushing/engulfment transition confirms this.

b) MORPHOLOGY OF THE SOLIDIFIED MATRIX

When an alloy solidifies in presence of particles, the diffusion of solute ahead of the solid-liquid interface is obstructed by the particles. This would lead to change in interface morphology and solute distribution in the solidified structure. This perturbation in the solute diffusion field decreases the solute concentration gradient between the particle and the interface. The thermal conductivity difference between the particle and the melt also alters the thermal gradient. The combination of these two effects contributes to the change in interface morphology observed in the present investigation. Thus, the observed changes in the microstructure (columnar to equiaxed-dendritic and cellular-dendritic) could be traced back to changes in solute and temperature fields brought about by the ceramic particulate.

c) SOLUTE BAND AROUND SiCp

The obstruction of the solute diffusion by the particle also leads to the accumulation of the solute in the region between the particle and the interface segment just below it. This slows down the movement of this part of the interface as compared to the rest. As the interface moves ahead and engulfs the particle, the accumulated solute remains trapped in the same region along with the particle. This explains the presence of a solute rich Cu band around the particle in the growth direction as shown in Fig.7. This possibility has also been suggested by Sekhar and Trivedi [8] in their studies on transparent organic systems.

d) INVERSE V-TYPE PARTICLE DEPLETED ZONE

In case of the permanent mold casting, the inverse 'V' type particle depleted zone is due to different cooling rates from the mold wall towards the central axis. Fig.10 schematically explains the cooling behaviour during permanent mold solidification. A plane near the mold wall (BB') experiences higher cooling compared to that at the centre (OO'), leading to a comparatively higher growth velocity near the mold wall. As discussed earlier, when the interface velocity is lower than the critical engulfment velocity, particles are pushed ahead of the advancing solid-liquid interface. As the pushing continues, more and more number of particles accumulate ahead of the interface forming clusters and with increasing cluster size the critical engulfment velocity continues to decrease (considering cluster size as the effective particle size). When the critical engulfment velocity drops below the actual interface velocity, the clusters get trapped. Since near the mold wall (along BB'), the actual interface velocity is higher, the critical cluster size is reached after a shorter distance of pushing, leading to a smaller particle free region. Whereas lower interface velocity towards the centre (OO') requires longer distance of traverse to form the critical cluster size and hence a larger particle depleted region.

e) GRAIN SIZE

The increase in grain size, found in case of Al-Cu-SiC$_p$ permanent mold casting, is on account of reduced convection currents due to the presence of SiC particles. It has been reported earlier that during solidification of MMCs in which reinforcing particles do not aid heterogeneous nucleation of the primary matrix phase, convection is suppressed [9]. Ohno [10] has discussed the mechanism of increase in grain size on account of reduced convection.

The secondary DAS increases away from the mold bottom on account of lowering of interface velocity. In presence of particles as the accumulation of particles ahead of the interface leads to coarsening and hence the higher DAS is observed in the composites as compared to the alloy.

f) HETEROGENEOUS NUCLEATION OF EUTECTIC SILICON ON SiCp

Heterogeneous nucleation of eutectic Si on SiC particles has been observed in the solidified castings. Not all of the eutectic Si seems to have nucleated on the SiC particles. The heterogeneous nucleation of primary Si on SiC is well documented [11,12]. Wang et al [11] have demonstrated that the nucleation of Si on SiC is more favourable on coherent or semi-coherent planes between the two crystal structures. They also showed that the semi-coherent relation between these two crystal systems with minimum mismatch is given by,

$$(1\ 1\ 1)_{Si}\ //\ (0\ 0\ 0\ 1)_{\alpha-SiC}$$
$$[0\ 1\ 1]_{Si}\ //\ [1\ 1\ 2\ 0]_{\alpha-SiC}$$

Since the SiC particles used in the present work have been produced by grinding, they possess a random orientation on their outer surfaces and this might prevent the availability of $(0\ 0\ 0\ 1)_{\alpha-SiC}$ planes for easier nucleation of Si crystals on SiC surface. Also, the oxidation of SiC and formation of spinels on SiC particles is expected to be non-uniform during the processing. Wang et al report that spinel layers on SiC particles assist Si nucleation on SiC. Non-uniformity of the spinel layer on SiC could be a possible reason behind non-uniform nucleation of Si on SiC particles. Presence of some free Si on the surface of SiC$_p$ could also aid the heterogeneous nucleation of eutectic Si on SiC$_p$.

CONCLUSIONS

1. Dendritic solidification experiments with Al alloys show complete failure of existing models. This can be attributed to the fact that all the theoretical models are based on the assumption that the solidification front is planar. But in dendritic solidification interface morphology is significantly different from that in the plane front solidification and it has been shown that this can change the behaviour of the forces acting on the particle.

2. Interaction between particle and the secondary dendritic arm has also been studied and it is observed that during growth some particles are engulfed by the secondary arms. Comparison of the experimental results with the theoretical model prediction shows that except the model of Shanguan et al., all other model agree with the observations.

3. Shape of the particle depleted zone has been explained qualitatively, in case of the permanent mold casting.

4. In case of unidirectional alloy solidification, presence of particles is found to change the nature of growth from columnar-dendritic to equiaxed-dendritic or cellular-dendritic. This can be attributed to the changes in the temperature and solute concentration gradients in the presence of particles.

5. The grain size in permanent mold castings is found to increase with increasing particle concentration. This could be attributed to reduced convective currents in the presence of particles. Secondary DAS is also found to be larger in the composite castings compared to that in the unreinforced alloy.

6. In A356-SiC composites, heterogeneous nucleation of eutectic Si on the SiC particles is observed but in Al-Cu and Al-Cu-Mg alloys, eutectic structure remains unaltered in the presence of particles.

7. Solute segregation around the SiC particle is observed in Al-Cu alloys. This is due to the physical barrier imposed by the particle to the diffusing solute flux.

ACKNOWLEDGMENTS

The authors wish to thank Department of Science and Technology for the financial support during the course of this study.

REFERENCES

1. D.R. Uhlmann, B. Chalmers and K.A. Jackson, Interaction between particles and a solid-liquid interface. *J. Appl. Phys.* **35**, *2986-2993* (1964).

2. G.F. Bolling and J. Cisse, A theory for the interaction of particles with a solidifying front. *J. Cryst. Growth* **10**, *56-66* (1971).

3. J. Postchke and V. Rogge, On the behaviour of foreign particles at an advancing solid-liquid interface. *J. Cryst. Growth* **94**, *726-* (1989).

4. R. Sasikumar and T.R. Ramamohan, Distortion of the temperature and solute concentration fields due to the presence of particles at the solidification front effects on particle pushing. *Acta. metall. mater.* **39**, *517-522* (1991).

5. D. Shanguan, S. Ahuja and D.H. Stefanescu, Behavior of ceramic particles at the solid-liquid metal interface in metal matrix composites. *Metall. Trans.* **23A**, *669-680* (1992).

6. A.M. Zubko, V.G. Lebanov and V.V. Nikonova, Reaction of foreign particles with a crystallization front. *Sov. Phys. Crystallogr.* **18**, *239-241* (1973).

7. M.K. Surappa and P.K. Rohatgi, Heat diffusivity criterion for the entrapment of particles by a moving solid-liquid interface. *J. Mat. Sci.* **16**, *765-767* (1981).

8. J.A. Sekhar and R. Trivedi, Solidification microstructure evolution in the presence of inert particles. *Mater. Sci. Eng.* **A147**, *9-21* (1991).

9. A. Mortensen and I. Jin, Solidification processing of metal matrix composites. *Int. Mater. Review* **37**, *101-128* (1992).

10. Atsumi Ohno, Solidification. The Separation Theory and it's Practical Applications Springer-Verlag, (1987).

11. W. Wang, F. Ajersch and J. P. Lofvander, Si phase nucleation on SiC particulate reinforcement in hypereutectic Al-Si alloy matrix. *Mater. Sci. Eng* **A187**, *65-75* (1994).

12. P.K. Rohatgi, S. Ray, R. Asthana and C. S. Narendranath, Interfaces in cast metal-matrix composites *Mater. Sci. Eng.* **A162**, *163-174* (1993).

APPENDIX

a_o	-	Interatomic distance
d	-	Gap thickness
g	-	Acceleration due to gravity
k	-	Boltzman constant
$\|m_l\|$	-	Liquidus slope
n	-	Constant = 5 for metallic materials
t_f	-	Local solidification time
B	-	Hamaker constant
$C\infty$	-	Solute concentration
D_l	-	Diffusivity
F_b	-	Buoyancy force
F_d	-	Drag force
F_N	-	Net force
F_r	-	Repulsive force
G	-	Temperature gradient
K_c	-	Partition coefficient
K_l	-	Thermal conductivity of liquid melt
K_p	-	Thermal conductivity of particle
L	-	Latent heat of fusion
N	-	Number of contact points
R	-	Particle radius
R_b	-	Bump radius
R_I	-	Interface radius
V_{cr}	-	Critical interface velocity
V	-	Interface growth velocity

ρ_p	-	Density of particle
ρ_l	-	Density of liquid
σ_{sl}	-	Solid-liquid interface energy
σ_{pl}	-	Particle-liquid interface energy
σ_{ps}	-	Particle-solid interface energy
$\Delta\sigma_o$	-	Interfacial energy difference
α	=	K_p/K_l
η	-	Viscosity of the melt
Ω	-	Atomic volume
$\psi(\alpha)$	-	Interface shape factor

LIST OF TABLES

TABLE 1 Details of castings prepared under different solidification conditions.
TABLE 2 Results of directional solidification experiments with copper chill at the bottom.
    (a) System : Al-4Cu-1vol% SiC_p
    (b) System : Al-4Cu-1Mg-1vol% SiC_p
TABLE 3 Models based on thermophysical properties.

LIST OF FIGURES

Figure 1 Particle engulfed by a secondary dendrite arm in directionally solidified Al-Cu-SiC_p composite.

Figure 2 Actual and theoretically predicted velocity profiles in the case of the permanent mold casting.

Figure 3 Macrostructure showing particle distribution in Al-Cu-SiC_p composite (a) directionally solidified and (b) permanent mold casting.

Figure 4 Dendritic morphology of the matrix in the particle free region and (b)cellular-dendritic matrix-morphology in the region of particle segregation in the directionally solidified Al-Cu-SiC_p composite.

Figure 5 Dendritic morphology of the matrix in the particle free region and (b)cellular-dendritic matrix-morphology in the region of particle segregation in the Al-Cu-SiC_p permanent mold castings.

Figure 6 (a) Variation of grain size as a function of SiC_p content and distance from the mold bottom in the permanent mold castings. (b) Variation of secondary DAS as a function of SiC_p content and distance from the mold bottom in the permanent mold castings.

Figure 7 Elemental mapping showing solute (Cu) trapped by the growing interface around the SiC particle in the directionally solidified Al-Cu-SiC_p composite.

Figure 8 SEM micrograph showing heterogeneous nucleation of eutectic Si on SiC_p.

Figure 9 Variation of forces acting on the particle with increase in interface velocity (a) in plane front solidification (b) dendritic solidification.

Figure 10 Schematic showing particle distribution and velocity profiles in the permanent mold casting.

TABLE 1. Details of castings prepared under different solidification conditions.

S.No.	MATRIX MATERIAL	SiC VOL% (SIZE IN μm)	SOLIDIFICATION CONDITIONS
1.	Al-3.2Cu	0 vol%	Permanent mold
2.	Al-3.2Cu	3 vol% (12μm)	Permanent mold
3.	Al-3.2Cu	5 vol% (12 μm)	Permanent mold
4.	Al-3.2Cu	10 vol% (12μm)	Permanent mold
5.	Al-4Cu	1 vol% (40μm)	Refractory molds (directional solidification)
6.	Al-4Cu-1Mg	1 vol% (40μm)	Refractory molds (directional solidification)
7.	A356	10 vol% (40 μm)	Refractory mold (directional solidification)

TABLE 2a Results of the Experiments with Copper Chill at the Bottom
System: Al-4Cu-1 vol.% SiC_p

Particle Size (μm)	Local Interface V ($\mu m/s$) Experimental Observation	U.C.J. V_{cr} ($\mu m/s$) Prediction	Cisse-Bolling V_{cr} ($\mu m/s$) Prediction	Shanguan V_{cr} ($\mu m/s$) Prediction	Postchke-Rogge V_{cr} ($\mu m/s$) Prediction	Sasikumar V_{cr} ($\mu m/s$) Prediction
25.5	0.35 Engulfed	0.0152 Engulfing	0.185 Engulfing	3.747×10^3 Pushing	4.67×10^{-5} Engulfing	0.207 Engulfing
28.0	0.248 Engulfed	0.0013 Engulfing	0.14 Engulfing	3.488×10^3 Pushing	4.67×10^{-5} Engulfing	0.198 Engulfing

TABLE 2b Results of the Experiments with Copper Chill at the Bottom
System: Al-4Cu-1Mg-1 vol.% SiC_p

Particle Size (μm)	Local Interface V ($\mu m/s$) Experimental Observation	U.C.J. V_{cr} ($\mu m/s$) Prediction	Cisse-Bolling V_{cr} ($\mu m/s$) Prediction	Shanguan V_{cr} ($\mu m/s$) Prediction	Postchke-Rogge V_{cr} ($\mu m/s$) Prediction	Sasikumar V_{cr} ($\mu m/s$) Prediction
6.5	5.38 Engulfed	0.234 Engulfing	11.2 Engulfing	1.47×10^4 Pushing	4.67×10^{-5} Engulfing	0.377 Engulfing
16.5	4.39 Engulfed	0.0363 Engulfing	0.684 Engulfing	5.92×10^3 Pushing	4.67×10^{-5} Engulfing	0.25 Engulfing

**TABLE 3. Results of the Experiments with copper chill at the bottom.
Models based on thermophysical properties**

Composite System	Particle Size μm	Local Interface Velocity ($\mu m/sec$).	Experimental Observation	Zubkov α Prediction	Surappa-Rohatgi β Prediction
Al-4Cu-1Mg -1 vol% SiC_p	6.5	5.38	Engulfed	0.3526 Pushing	0.4436 Pushing
Al-4Cu -1 vol% SiC_p	25.5	0.35	Engulfed	0.3526 Pushing	0.4436 Pushing

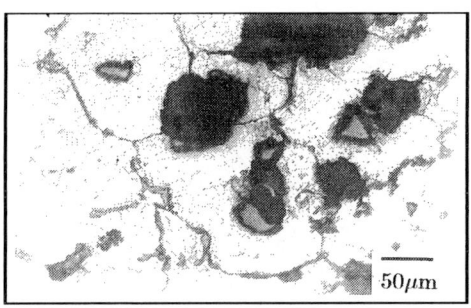

Figure 1 Particle engulfed by a secondary dendrite arm in directionally solidified Al-Cu-SiC_p composite.

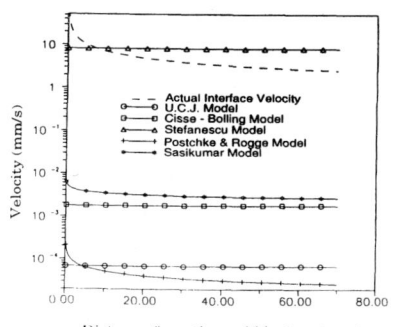

Figure 2 Actual and theoretically predicted velocity profiles in the case of the permanent mold casting.

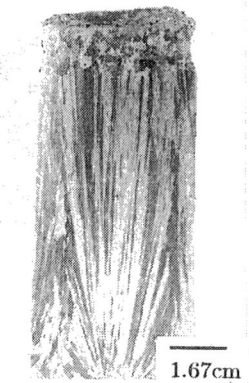 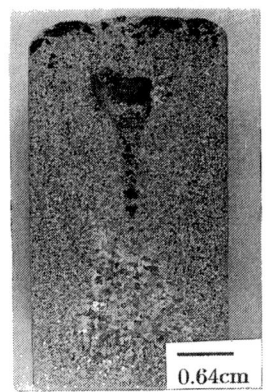

Figure 3 Macrostructure showing particle distribution in Al-Cu-SiC$_p$ composite (a) directionally solidified and (b) permanent mold casting.

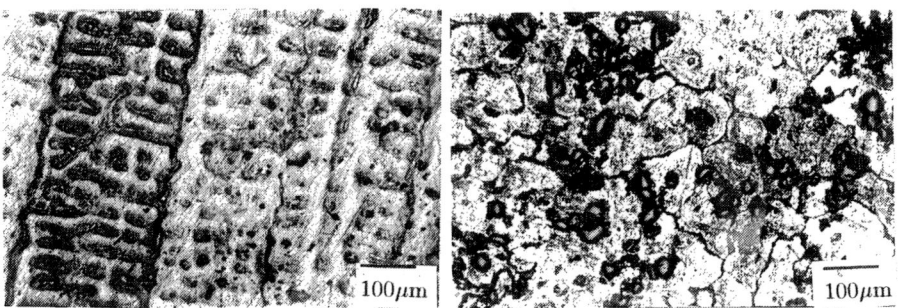

Figure 4 Dendritic morphology of the matrix in the particle free region and (b)cellular-dendritic matrix-morphology in the region of particle segregation in the directionally solidified Al-Cu-SiC$_p$ composite

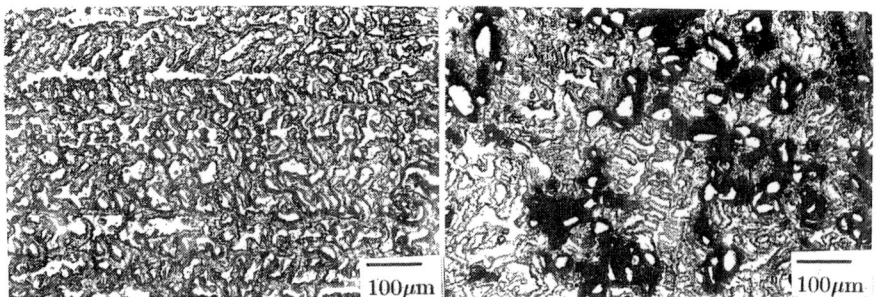

Figure 5 Dendritic morphology of the matrix in the particle free region and (b)cellular-dendritic matrix-morphology in the region of particle segregation in the Al-Cu-SiC$_p$ permanent mold castings.

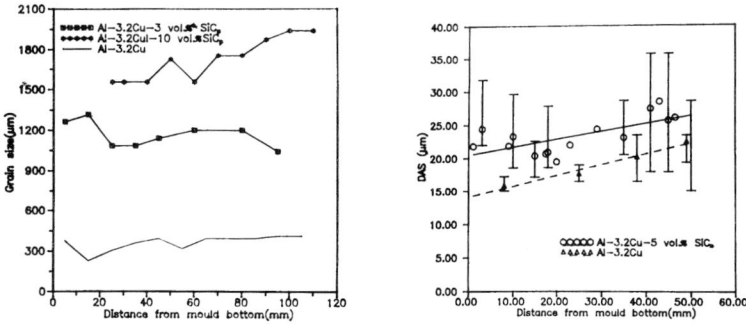

Figure 6 (a) Variation of grain size as a function of SiC_p content and distance from the mold bottom in the permanent mold castings. (b) Variation of secondary DAS as a function of SiC_p content and distance from the mold bottom in the permanent mold castings.

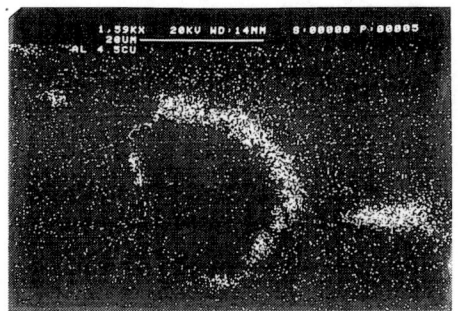

Figure 7 Elemental mapping showing solute (Cu) trapped by the growing interface around the SiC particle in the directionally solidified Al-Cu-SiC_p composite.

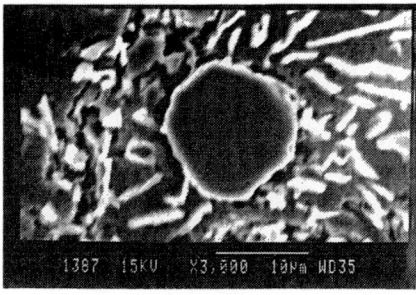

Figure 8 SEM micrograph showing heterogeneous nucleation of eutectic Si on SiC_p.

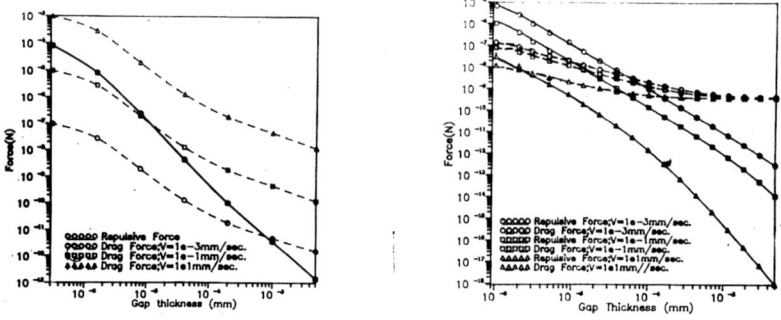

Figure 9 Variation of forces acting on the particle with increase in interface velocity (a) in plane front solidification (b) dendritic solidification.

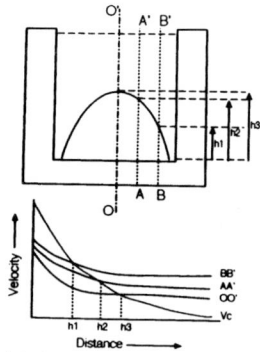

Figure 10 Schematic showing particle distribution and velocity profiles in the permanent mold casting.

Fabrication and mechanical properties of *in situ* formed carbide particulate reinforced aluminum composite.

TAKAO CHOH

Department of Materials Processing Engineering, School of Engineering, Nagoya University, Furo-cho, Chikusa-ku, 464 Nagoya, Japan.

Abstract

Stable carbide particles of TiC, ZrC and TaC were *in situ* synthesized in liquid aluminum by the reaction between Al-Ti, Al-Zr or Al-Ta system liquid alloy and SiC or Al_4C_3 particles at 1473K. However, the dispersion behavior of TiC particles in the matrix depended on the size of the raw carbide. Finer SiC made the dispersion of TiC particles more uniform, resulting in more improving the mechanical properties. Furthermore, although Al-Ti-Si system intermetallic compound was detected in a TiC_p/Al-Si composite fabricated by the melt stirring method, those compounds considerably decreased in the composite fabricated by the *in situ* method. The mechanical properties of *in situ* formed TiC_p/Al-5wt%Mg and TiC_p/Al-5wt%Cu composites were better than those fabricated by the melt stirring method, and by T6 heat treatment, those properties of *in situ* formed TiC_n/Al-5wt%Cu composite were further improved. The experimental results were analyzed by the reaction model based on the assumption that the overall reaction rate was controlled by both the reaction and the diffusion.

1. Introduction

In recent years, much of attention has been given to the development of effective fabrication process of metal matrix composites. In the *in situ* fabrication process, the spontaneous reaction between the raw materials is utilized to synthesize reinforcements in the metal matrix. Thus, it is expected that the *in situ* formed composites may reveal not only excellent dispersion of fine reinforcing particles, but also high thermodynamical stability.

P.Sahoo and M.J.Koczak[1] reported that the *in situ* formed TiC_p/Al composites were fabricated by bubbling carbonaceous gas into the Al-Ti alloy melt and revealed excellent mechanical properties. However, the process using carbonaceous gas has some practical difficulties, such as the requirement of complicated equipment and problem in controlling the volume fraction of formed TiC particle. Therefore, a simplified process is looked forward to solving these problems.

This work dealt with the fabrication of *in situ* formed carbide particulate reinforced aluminum composites by utilizing the reaction between liquid aluminum alloy containing thermodynamically stable carbide formation element and relatively unstable carbide such as SiC and Al_4C_3 as the solid carbon source. The dispersion behavior of particle formed by *in situ* reaction process, the mechanical properties of *in situ* composites and the mechanism of *in situ* formation of particles in the liquid aluminum alloy were investigated.

2. Experimental Procedure

In the first method, 80g of aluminum containing carbide formation element of titanium, zirconium and tantalum was melted in an induction furnace in MgO crucible under an argon atmosphere, and held at 1423K. Then, SiC or Al_4C_3 particles were added and incorporated by the melt stirring method.[2] During this stirring, TiC, ZrC and TaC particles were formed *in situ* in the liquid aluminum and subsequently the melt was cast into a permanent mould. In part of experiment, some samples of about 3g were taken from melt using an alumina-coated quartz tube during stirring in order to investigate the time dependence of the *in situ* reaction.

In the second method, the powder mixture of titanium and Al_4C_3 particles was intermittently added to pure aluminum at the rate of about 1g/30sec at 1473K during stirring. After incorporation of all of added particles, melt stirring was continued for 300s to complete the *in situ* reaction. The melt temperature was lowered to 973K and an alloying element of magnesium or copper was added to produce Al-5wt%Mg or Al-5wt%Cu matrix. Subsequently those samples were solidified in the crucible by air cooling outside furnace.

The samples were extruded at 773K with a extrusion ratio of 25. Part of each specimen was machined, to investigate the mechanical properties, the other part was polished for microstructural studies by SEM, EPMA and XRD. SiC_p (nearly 14 μm in diameter) /Al-Si and TiC_p (nearly 1μm)/Al-Si composites were also fabricated by the melt stirring method at 1073K with an appropriate stirring time, in order to compare with the *in situ* fabricated composites. The theoretical volume fraction of *in situ* formed particles reaches 5.5% in the first method, and 5 and 10% in the second method, assuming complete *in situ* reaction.

3. Results and discussion

3.1 Fabrication and mechanical properties of *in situ* formed TiC particulate reinforced aluminum composite by the first method

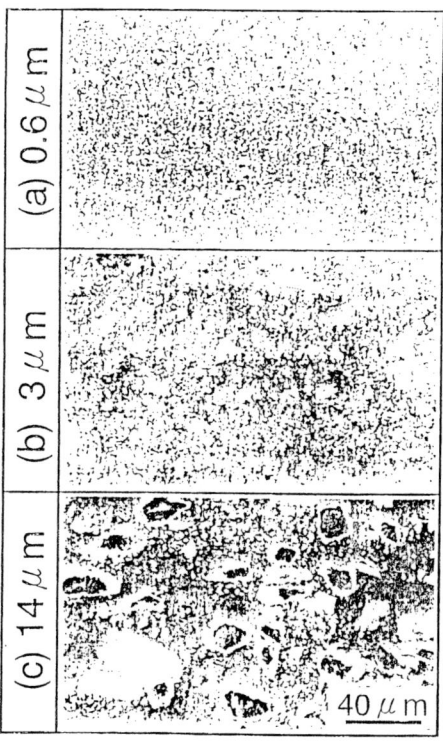

Although SiC particles are poorly incorporated into the melt with decreasing particulate diameter[3], by elevating melt temperature to 1473K, fine SiC particles as small as 0.6 μm diameter could be incorporated into the melt, because of the essentially improved wettability.

The microstructure of *in situ* composites fabricated are compared in Fig.1. It is obvious from Fig.1(a) that approximately 1 μm particles are almost uniformly dispersed in the composites fabricated by adding 0.6 μm SiC. The same distribution of particles can be achieved by using 3 μm SiC as in Fig.1(b). However, using 14 μm SiC particles shown in Fig.1(c), the circular TiC phase is observed surrounding the remained raw SiC particles as well as the large intermetallic compounds of Al_3Ti. Hence, the size of SiC particle has an effect on the rate of the *in situ* reaction, resulting in heterogeneous dispersion behavior. Further, ultimate tensile strength, 0.2% proof stress and uniform elongation of

Fig.1 Scanning electron micrographs of *in situ* TiC_p/Al composites fabricated by reaction time 900 sec at 1473K from (a) 0.6 μm, (b) 3 μm and (c) 14 μm SiC.

those composites were decreased with increasing diameter of SiC as shown in Fig.2, depending on such particle dispersion behavior.

The mechanical properties of *in situ* formed TiC_p/Al-5wt%Si composite are compared with those composites reinforced with nearly 14 μm SiC and nearly 1 μm TiC particles fabricated by the melt stirring method, as shown in Fig.3. It is found that ultimate tensile strength and 0.2% proof stress of *in situ* TiC_p/Al-5wt%Si composite are higher than those of two others, namely strength efficiency is improved through the *in situ* reaction.

On observing the microstructures of TiC_p/Al-5wt%Si composites fabricated both (a) by the melt stirring method and (b) the *in situ* method, it should be noted that large precipitates are found in the matrix in the case of melt stirring method as shown in Fig.4, though the particle size of TiC in both composites is not markedly different. From EDX analysis, those precipitates were determined as Al-Ti-Si system intermetallic compound. This compound may be mainly precipitated through the following process as shown in Fig.5 where titanium and carbon contents in (a) conventional melt stirring method are initially increased by the dissolution of added TiC particle, then the titanium content reaches its solubility ($3Al + Ti = Al_3Ti$) before the achievement of equilibrium between the titanium content and the carbon content in the liquid aluminum, further the precipitation of Al_3Ti proceeds till the carbon content in the melt increases to the equilibrium value with the above mentioned titanium solubility content.

On the other hand, in the case of (b) the *in situ* composite, those compounds decreased remarkably. It suggests that main part of titanium atoms combine with carbon supplied from SiC, and then carbon and titanium contents may approach to equilibrium value, and during cooling, TiC particles are formed, resulting in a decreasing precipitation of Al_3Ti. Thus, the *in situ* composite free from the precipitation of Al_3Ti can be fabricated, if the ratio of titanium and carbon content supplied in the melt is adequate.

3.2 Mechanical properties of *in situ* formed ZrC and TaC particulate reinforced aluminum composite

In situ formed ZrC_p/Al and TaC_p/Al composites were fabricated by the first method, adding 14 μm Al_4C_3 particles to Al-Zr and Al-Ta liquid aluminum alloys and stirring for 1200 sec at 1473K. Fig.6 shows the mechanical properties of fabricated composites, comparing with those of *in situ* formed TiC_p/Al composite.

Ultimate tensile strength and 0.2% proof stress of *in situ* TiC_p/Al composite were highest, although the highest uniform elongation was obtained in the *in situ* TaC_p/Al composite.

3.3 The effects of magnesium and copper on the mechanical properties of *in situ* formed TiC_p/Al composite

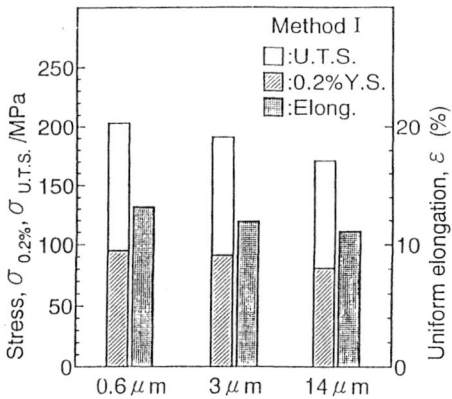

Fig.2 Mechanical properties of in situ TiC$_p$/Al composites fabricated by reaction time 900 sec at 1473K from 0.6 μm, 3 μm and 14 μm SiC.

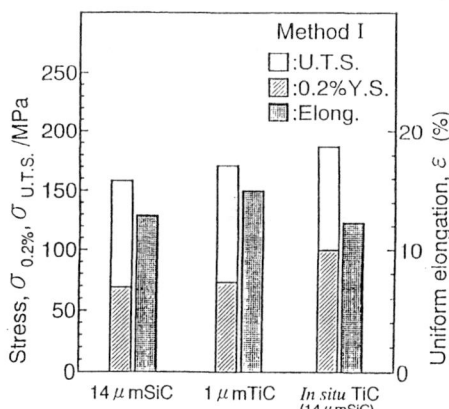

Fig.3 Comparison of mechanical properties of in situ TiC$_p$/Al-Si composites fabricated from 14 μm SiC with those of 14 μm SiC$_p$/Al-Si and 1 μm TiC$_p$/Al-Si composites fabricated by the melt stirring method.

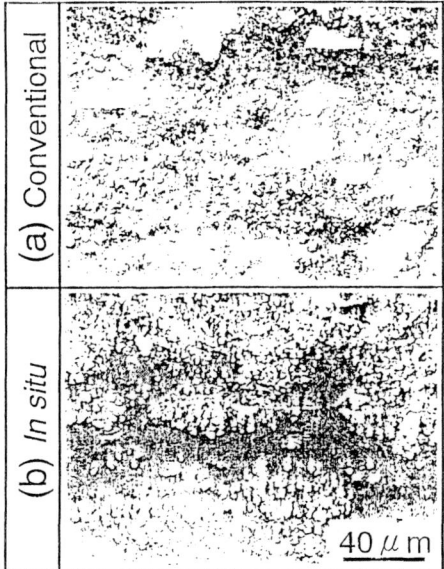

Fig.4 Scanning electron micrographs of TiC$_p$/Al-5wt%Si composites fabricated by (a) the melt stirring method and (b) the in situ method.

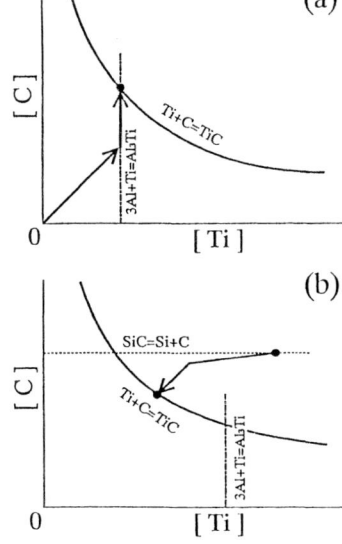

Fig.5 Ideal relationships between carbon and titanium contents equilibrated with titaniumu carbide in liquid aluminum.
(a) conventional melt stirring (1073K)
(b) in situ processing (1673K)

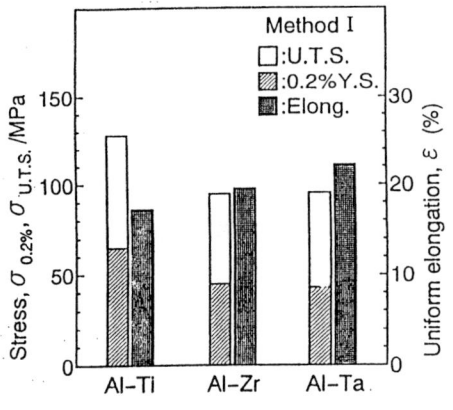

Fig.6 Mechanical properties of *in situ* composites fabricated through the reaction between 14 μm Al_4C, particle and Al-Ti, Al-Zr and Al-Ta system liquid alloys with reaction time 1200 sec at 1473K.

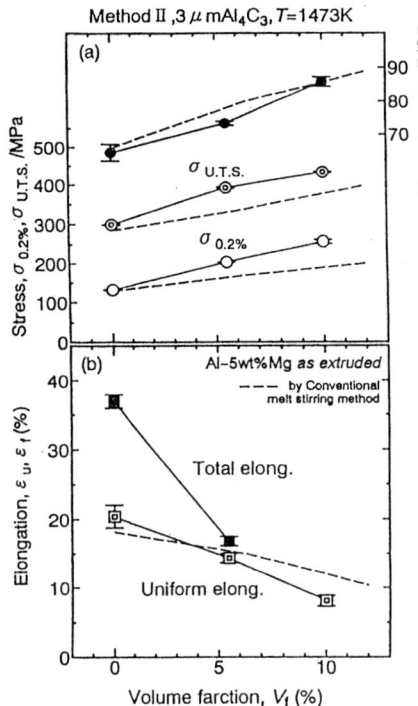

Fig.7 Comparison of mechanical properties of in situ TiC_p/Al-5wt%Mg composites with that fabricated by the melt stirring method[4].

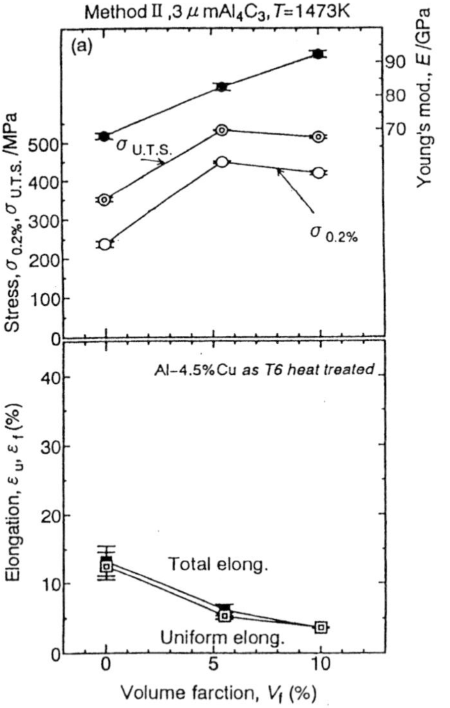

Fig.8 Effect of the particulate volume fraction on the mechanical properties of *in situ* TiC_p/Al-4.5wt%Cu composite after T6 heat treatment.

In situ formed TiC$_p$/Al-5wt%Mg and TiC$_p$/Al-5wt%Cu composites were fabricated by the second method, where a powder mixture of 3 μm Al$_4$C$_3$ and 99%Ti (neary 350 mesh) was added to the melt at 1473K in order to increase the volume fraction of the *in situ* formed particulate reinforcement. After the *in situ* reaction had ceased, magnesium or copper was alloyed at lowered temperature. The values of Young's modulus, ultimate tensile strength and 0.2% proof stress of *in situ* TiC$_p$/Al-5wt%Mg composite were increased with increasing volume fraction of *in situ* formed TiC particles, as shown in Fig.7. Furthermore, it is noticeable that the ultimate tensile strength and 0.2% proof stress are higher than those obtained in the same composite fabricated by the melt stirring method, as described by dotted lines.[4] This is contributed to by the better distribution of *in situ* formed TiC particle in the matrix without the agglomeration of particle. However, the uniform elongation decreased with increasing volume fraction of TiCp, and, above nearly 10% of volume fraction, became rather less than that by melt stirring method.

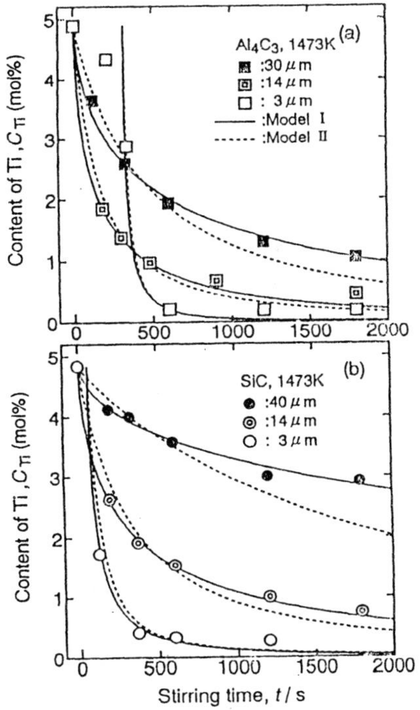

Fig.9 Effect of the initial diameter of raw carbides on the in situ reaction rates in (a)Al$_4$C$_3$/Al-Ti alloy and (b)SiC/Al-Ti alloy systems, and the applications of model 1 and model 2.

On the other hand, Young's modulus, 0.2% proof stress and ultimate tensile strength of the *in situ* TiC$_p$/Al-4.5wt%Cu composite after a T6 heat treatment increased and the elongation decreased with increasing volume fraction of TiC$_p$, as shown in Fig.8.

3.4 Mechanism of *in situ* formation of particle

It was found that the *in situ* reaction proceeded leaving the *in situ* formed TiC particles crowded together in a layer around the raw materials of Al$_4$C$_3$ or SiC particles. This suggests that titanium atoms must diffuse in this layer to react with the raw materials. This reaction type is similar to the gaseous reduction of iron oxide pellets that has been generally analyzed by the so-called unreacted core model.[5]

Hence, the process of *in situ* formation of particles was analyzed by applying the modified unreacted core model, assuming that both *in situ* chemical reaction and titanium diffusion contribute to the overall reaction rate.

Eq.(1) can be obtained finally.

$$(1/K_1)\{(C/C_0)^{-2/3}-1\} + (1/K_2)\{\ln(C/C_0) + 3(C/C_0)^{-1/3} - 3\} = C_0 t \tag{1}$$

where Z_0: diameters of raw carbide at initial(m), C_b, C_{b0}: titanium content and initial one in the melt bulk(mol/m^3), ρ_{Al} is density of aluminum (kg/m^3) and M_{Al} is atomic weight of aluminum(kg/mol), $f = \rho_{Al}/(100 \cdot M_{Al})$, $C_b = fC$ and $C_{b0} = fC_0$. K_1 and K_2 are given by as follows:

$$K_1 = (4fk_R/\delta_0)(1/Z_0) \tag{2}$$
$$K_2 = (6fD/\delta_0)(1/Z_0)^2 \tag{3}$$

The first and the second terms in the left-hand side of Eq.(1) respectively express the contributions of chemical reaction and diffusion. Therefore, assuming low resistance of diffusion, as in the case of a small raw carbide, Eq.(1) becomes Eq.(4).

$$(1/K_1)\{(C/C_0)^{-2/3} - 1\} = C_0 t \tag{4}$$

Now, Eqs.(1) and (4) are applied to the experimental results as the model 1 and the model 2. Fig.9 shows the time dependence of the titanium content in the melt, analyzed by the ICP emission spectrometric analysis method. The rate of consumption of titanium at 1473K by the raw carbide of Al_4C_3 or SiC increases with decreasing diameter of those carbides. However, in the cases of 3 μm Al_4C_3 and 3 μm SiC, the rates of the initial stage are decreased by poor wettability due to the smaller diameter.[3]

In order to apply the model, the values of K_1 and K_2 were estimated from the values of the intercept and the slop in the plotted relationship between $t/\{(C/C_0)^{-2/3}-1\}$ and $\{\ln(C/C_0)+3(C/C_0)^{-1/3}-3\}/\{(C/C_0)^{-2/3}-1\}$ obtained by deviding both sides of Eq.(1) by $\{(C/C_0)^{-2/3} - 1\}$. On the other hand, the value of K_1 in model 2 was obtained from the slope of the linear relationship between $C_0 t$ and $\{(C/C_0)^{-2/3}-1\}$. The solid and dotted lines in Fig.9 show the applied results of Eqs.(1) and (4) by using already estimated values of K_1 and K_2. The solid lines well agree with the experimental results, compared with the dotted lines of model: namely the contribution of diffusion resistance in the *in situ* formed TiC particle agglomeration layer can not be ignored.

Detail in this work may be given elsewhere[7]

4. Conclusion

In situ formed carbide particulate reinforced aluminum composites were fabricated by using SiC and Al_4C_3. The mechanical properties, the dispersion behavior of *in situ* formed particles, and the mechanism of *in situ* formation of particle were investigated.

1. Fine TiC particles of 1 μm in diameter were *in situ* formed even by using 14μm SiC. However, the finer the raw carbide of SiC and Al_4C_3 were, the more the *in situ* formed particle dispersed uniformly and the more the mechanical properties of composite were improved.
2. The mechanical properties of composite fabricated from Al_4C_3/Al-Ti system were better than those obtained from Al_4C_3/Al-Zr and Al_4C_3/Al-Ta systems.
3. The mechanical properties of *in situ* formed TiC_p/Al-5wt%Mg composite and the *in situ* formed TiC_p/Al-5wt%Cu composite after T6 heat treatment were better than those fabricated by the melt stirring method.
4. It seems reasonable to consider that the rate of *in situ* carbide formation could be analyzed by the mixed control of the interfacial chemical reaction and the titanium diffusion in the *in situ* formed TiC particles agglomeration layer.

References

1. P.Sahoo and M.J.Koczak: Proc.1st Japan Intern. SAMPE Sympo., (1989),958
2. M.Kobashi and T.Choh: J. Japan Inst. Met., 55(1991)731
3. T.Choh, Z.Ebihara and T.Oki: J.Japan Inst. Light Met., 39(1989)356
4. M.Kobashi, M.Harata and T.Choh: J.Japan Inst. Light Met., 43(1993)522
5. For example,K.Mori: Tetsu-to-Hagane, 50(1964)2259
6. M.Kobashi and T.Choh: J.Japan Inst.Met., 55(1991)79
7. H.Nakata, N.Kanetake and T.Choh: J. Mater. Sci., to be published.

The Development of Cast Metal Matrix Silicon Carbide Particle Reinforced Composites

B. Inem
Gazi University, Faculty of Technical Education, Metallurgy Department, Ankara, Turkey

Development and microstructure/property relationships of new commerical sand cast (ZC63) or ZCM630 (Mg-6%Zn-3%Cu-0.5%Mn) and wrought (ZC71) or ZCM711 (Mg-7%Zn-1%Cu-1%Mn) magnesium alloy composites reinforced with up to 20% β-SiC particles, with average diameter of 15 μm, have been investigated. The optimum particle volume fraction for the cast composites was established. A significant change in the grain size of the composites was observed. The effect of the SiC particles on actual grain size has been discussed and quantified. As well as increasing the hardness of composites the SiC particles accelerated the age hardening response. The rapid ageing kinetics in composites has been related to a higher diffusion rate which results from dense dislocation networks which often appeared around the SiC particles. Analytical electron microscopy showed a featureless interface in all composites although preferential nucleation of the eutectic phase on the SiC was observed.

The tensile strength and dynamic moduli of the composites were measured and compared with those of the unreinforced alloys. An increase of about 19% in tensile strength and 33% in modulus were found and the latter result is discussed in terms of the relationship between particles and the eutectic and bonding at both the eutectic/particle and matrix/particle interface.

The thermal expansion coefficient of both the die cast and extruded ZCM711 was found to be close to the lower limit of existing models, whereas the thermal expansion coefficient of ZCM630 was found to fit a upper limit (Rule of Mixtures) relationship. These results are attributed to the microstructural defects and the distribution of the particles throughout the composites body, as well as interface bonding.

1. Introduction

Over the past five or six years much of the development in metal matrix composites (MMCs) has been in the use of particle reinforced composites produced by powder metallurgy (PM), and towards liquid processing techniques (LPT). Production of MMCs by LPT have been found to be the most promising for the manufacture of near-net shape components at a relatively low cost.

The use of magnesium alloys as the matrix phase in the MMCs research has been of interest and they have been considered as an alternative to aluminium base composites for advanced structural applications and for components in engines with the advantage of high specific strength and stiffness. To date limited studies has reported on the cast particle reinforced magnesium matrix composites(1-4).

The microstructures of composites have an important determinant effect on the properties of materials, including the dynamic modulus and the coefficient of thermal expansion

Inorganic Matrix Composites
Edited by M. K. Surappa
The Minerals, Metals & Materials Society, 1996

(CTE). The aim of present study has been to examine and relate the microstructures and properties in the newly developed zinc, copper, manganese containing sand cast (SC) ZCM630 and die cast (DC)/ extruded ZCM711 magnesium matrix composites, reinforced with SiC particles.

2. Experimental Procedures

The materials used for present study were a magnesium cast alloy (ZCM630) with a nominal composition Mg-%6Zn, %3Cu, %0.5Mn and a wrought alloy (ZCM711) having a nominal composition Mg-%7Zn, %1Cu, %1Mn, and commercial purity magnesium (CPMg) for comparison provided by Magnesium Elektron Limited (MEL), U.K. The cast alloy was reinforced with up to 20% β-SiC particles and the wrought alloy was reinforced with 12% β-SiC particles having an average diameter of 15μm. The ZCM630 composite was cast in the form of plates (250x250x25 mm) and the ZCM711 composite was cast in a water cooled die with a diameter of 80mm and subsequently extruded by a % 80 reduction in diameter.

The samples were examined in both as cast and heat treated condition. ZCM630, ZCM711 samples unreinforced and composite were solution heat treated at 440 °C for 8h in a protective atmosphere of argon in a closed alumina tube furnace followed by a hot water (65 °C) quench. After quenching, the samples were aged at 200 °C and cooled in air. Subsequently Vickers hardness measurements were carried out. At least 10 measurements were made from each sample. The elastic moduli of samples were measured using a sonic moduli measurement technique. CET of both matrices and composites were measured using a thermomechanical analyser (TMA) at temperature range of 20-350 °C.

For light (LM) and scanning electron microscopy (SEM), the samples were ground through 1200 grit SiC abrasive and polished on a soft cloth using 6μm, 3μm and 1μm diamond paste and etched in tartaric acid, nital or acetic picral solution.

ZCM630 and ZCM711 composite samples in the as-cast, solution heat treated and aged condition were examined using Joel 200CX and 2010 transmission electron microscopes (TEM) at 200 kV. Foils for TEM were prepared as follows: after electro discharge sectioning the samples were ground to approximately 60μm and subsequently ion beam thinned at 5 kV, 0.4 mA and 30°,15°,6° impingement angle.

3. Results and Discussion

3.1. Solidification Microstructure ZCM630 15%SiC SC Composite

LM and SEM studies showed the distribution of SiC particles throughout the structure to be uniform at low magnification. However with a higher magnification, in ZCM630 composite particles segregated in the interdendritic regions (Fig.1). This indicates that during solidification of the composite, SiC particles were pushed by the primary magnesium grains into the last solidifying interdendritic regions and mostly trapped between dendrite arms or in the grain boundaries of the matrix. Similar results were reported in previous publications[1-3,4]. A number of particles were however captured within the grains.

Agglomerates of particles have been seen in the cast composite. Pores and other casting defects in the composite have also been observed. This is due to difficulty in dispersing fine SiC particles in liquid magnesium and wettability of SiC during the composite consolidation.

With 10% SiC particles in ZCM630 matrix SC composite, particle settling was observed leaving almost 5 mm particles free region at the top of the cast plate. With a higher volume fraction of particles, on the other hand, slight settling of particles occurred leading to a little increase in the particle density at the bottom, but almost no particle free region at the top of the casting.

Although the wettability of SiC by liquid Mg is better than that by liquid Al; as with an Al matrix composite SiC particles were rejected by the molten Mg. However with the Mg alloy the final solidifying eutectic phase $Mg(Cu\ Zn)_2$, appeared to be dispersed around the particles. The lamellar eutectic nucleated from the SiC surface[5]. After heat treatment the lamellar eutectic spheroidized on the surface and around particles in the grain boundary regions.

Clustering of the SiC particles was frequently associated with oxide. It seems that during the injection of particles and stirring of the molten metal to disperse the particles, gas and air on the particles surface diffused into the molten metal and caused oxide formation which clustered the particles and interrupted the homogeneous distribution of particles. A similar observation has been reported by other workers[3].

Some hot tearing was observed in both composites (Fig. 2). These tears also appeared at the fracture surface as a intergranular crack. Although unreinforced ZCM630 and ZCM711 matrix alloys are not susceptible to hot tearing, the problem has been experienced with the addition of reinforcing particles and is explained elsewhere[5].

3.2. Microstructure of DC and Extruded ZCM711 with 12%SiC Composite

With a relatively high solidification rate and modified composite consolidation, the defects such as oxides, porosity, agglomeration of particles were reduced in the DC composite. As it was expected, no hot tearing took place. Because increased nucleation rate due to water cooling of the die resulted in a fine grain structure, fewer particles are trapped at the grain boundaries than in SC composites. The eutectic phase in this case also solidified at the surface of the SiC particles.

In a DC billet a 2mm wide particle free region was observed at the edge of the billet where the cooling rate is a maximum(Fig.3). This indicates that on the rejection of particles by primary matrix grains causes the formation of particle free region. There is a critical solidification rate below which the gravity effective is on the particle settling as well as volume fraction of particles in the composites[6], and above which, due to the water cooled die wall high nucleation and growth rate of the grains, particles pushing results in a particle free region.

During the composite extrusion, however, most of the casting defects were eliminated. Although partial alignment of SiC particles in the extrusion direction occurred, a homogeneous distribution of SiC particles throughout the structure was produced (Fig.7). The lamellar eutectic phase around the particles was mostly broken and spheroidized and dispersed throughout the structure. A small amount of eutectic remained around the particles.

TEM revealed that dislocations were punched into the matrix by the SiC particles and tangled around the particles. This indicates the existence of a stress field which was induced by the thermal expansion mismatch. In most cases a featureless SiC/matrix and SiC/eutectic interface is observed. However, occasionally at the surface of the SiC particles a needle-like Mg_2Si reaction product has been observed. This has been discussed in previous publications[5,7,8]. The tendency for the formation of such a reaction product increased with the increasing composite process time at high temperature.

3.3. Grain Size and Morphology of Composites

The distribution of particles in the melt altered the structure of the composites, influencing the solidification behaviour in ZCM630 SC composite and resulting in a fine grain structure. In ZCM711 DC composites, both the particles and rapid solidification also gave rise to a much finer grain size, although there is no direct indication that the SiC particles act as nuclei for the magnesium grains. If this was the case, first more particles would remain in the magnesium grains and not at the grain boundaries and second the last solidifying eutectic would mostly not have nucleated from the particles, because the particles would have been surrounded by the magnesium grains. The SiC particles, however, blocking further grain growth during solidification resulted in a fine grain structure.

The grain sizes of the composites and unreinforced alloys examined are given Table 1. There is a clear difference between the grain size of composites and the grain size of unreinforced alloys. However, the grain size in both the SC and DC composites varied throughout the composites. This difference in grain size is a function of solidification rate, the local volume fraction of reinforcement and the population of particles around the grains. It is evident that in the some regions where the cooling rate is slow, the grain size increases as a result of decreasing nucleation rate, with an extensive grain growth causing more particle collecting at the grain boundary and an increasing tendency for particle trapping. Figure 8 shows the effect of the particles on grain size. As the volume fraction of particles increases the grain size in the composites decreases.

Table 1: The grain size of composites and matrix alloys

Samples	Grain Size (μm)	
	Before recrystallization	After recrystallization
ZCM630 alloy	142	No recrys.
ZCM630+10%SiC	113	41
ZCM630+15%SiC	86	37
ZCM711 DC	70	No recrys.
ZCM711+12%SiC	50	30
ZCM711 Extruded	70 (after DC)	21 (Dynamic recrys)
ZCM711+12%SiC	50 (after DC)	10 (Dynamic recrys)

3.4. Recrystallization in As Cast and Extruded Composites

The other factor for further grain refinement is recrystallization in both as cast and extruded composites. It is well established[9,10] that extensive deformation around the particles in extruded aluminium matrix composites give rise to preferential nucleation side for recrystallization resulting in a much finer grain size compared with that of unreinforced extruded matrix alloy. It has been quantified in ZCM711 12% SiC extruded composites that in most cases more than six magnesium grains nucleated on the SiC particles during the dynamic recrystallization reducing the grain size of composites to less than 12 μm after a 80% extrusion[11].

In the cast alloy and CPMg (Commerical Purity Magnesium) composites with an identical recrystallization heat treatment, as can be seen in (Fig.9), static recrystallization effectively around the SiC particles has occurred. This result has also been confirmed by a PEEM (Photo Emission Electron Microscopy) study[5]. On the other hand, in the unreinforced matrix samples with the same heat treatment cycle the recrystallization did not take place. It is believed that in the cast composite thermal expansion mismatch induced internal stresses, high dislocation densities and heavy twinning all of which are responsible for such static recrystallization.

3.5. Ageing Response and Hardness of Composites

In all composites, the addition of SiC particles caused an increase in hardness and accelerated the ageing response of the composites compared to the their unreinforced matrices. Table 2 shows that in the case of alloy composites in the fully heat treated condition, a 15% increase for ZCM711+12% SiC extruded, a 25% increase for ZCM711+12% SiC DC and a 40% increase for ZCM630+15%SiC SC composite were observed. This improvement in hardness is due to fine grain size of the matrices in the composites, as well as presence of the particles.

Fig.10a, 10b, 10c, shows the ageing response of composites. ZCM630+15% SiC SC and ZCM711+12% SiC extruded composites demonstrate a rapid ageing and both composites reached the peak hardening at the same time with their unreinforced matrices. However, ZCM711+12% SiC DC composites reached to the peak hardness in a shorter exposure time than that of the unreinforced DC matrix. This indicates that in DC cast composites the ageing kinetics are faster than those two composites. However, a previous study on ageing of cast SiC reinforced Mg-6Zn alloy showed the behaviour to be similar to that observed in the unreinforced alloy[12]. In this case the result has been related to the absence of a SiC particles/matrix interface associated dislocation substructure. It is mentioned above that the thermal expansion mismatch between the matrix and the reinforcement caused to the stress field around the particles during the solidification and quenching after the solution heat treatment. This also produced extensive twinning and an increase in dislocation density (Fig.11). Such dislocations and twinning have been observed even after extrusion in which recrystallization and annealing mostly eliminates the residual stress in the metal and alloys. However, in the case of metal/ceramic composites, in any state such as after annealing, slow cooling the residual stress and increased dislocation density around the particles the most likely exist due to the large thermal expansion differences between metal matrix and ceramic reinforcement. if an adequate interface bonding exist and no debonding occurs during the annealing. Therefore, the increase in dislocation density produced more nucleation sites for both β' and β'' precipitates resulting in fine and dense precipitates around the particles in the stress field (Fig.12). Because while β' precipitates nucleate on screw dislocations, β" precipitates nucleate on edge dislocation in both ZCM630 and ZCM711 alloys.

Table 2: Hardness of composites and unreinforced alloys

Materials		Hardness (Hv)
ZCM630	(F)	55
ZCM630	(T6)	60-72
ZCM630+10%SiC	(F)	57
ZCM630+10%SiC	(T6)	69
ZCM630+15%SiC	(F)	79
ZCM630+15%SiC	(T6)	101
ZCM711 DC	(F)	54
ZCM711 DC	(T6)	73
ZCM711+12%SiC DC	(F)	84
ZCM711+12%SiC DC	(T6)	91
ZCM711 Extruded	(F)	65
ZCM711 Extruded	(T6)	76
ZCM711+12%SiC Ext	(F)	85
ZCM711+12%SiC Ext	(T6)	90

3.6. Dynamic Elastic Modulus of Composites

Table 3 shows the elastic modulus results for composites and unreinforced matrices. The modulus of all composites increased with the addition of SiC particles. The increase is 33% for the ZCM630 matrix composite with 15% SiC particle content. It is believed that in alloy composites, solidification of the lamellar eutectic $\{Mg(CuZn)_2\}$ at the SiC particles surface[8] produced a higher modulus.

The increases in moduli for DC and extruded ZCM711+12% SiC composites, on the other hand, were 20% and 24% respectively. The increase in modulus for extruded composites was found to be slightly higher than the increase in moduli for the DC composite. This indicates that pores and probably other casting defects which are present in the DC billet could be responsible for the relatively lower modulus value in this composite compared with that of extruded composite in which such casting defects were mostly eliminated. Voids and pores in the bulk materials behave as a third phase with zero moduli. Therefore, according to ROM (Role of Mixture) the simplistic approach to estimate modulus, as the volume of casting defects decreases, the modulus of composites increases. Thus, if the casting defects in the ZCM630+15%SiC SC composite are reduced, a further increase in moduli of this composite would be possible.

Table 3: Modulus of composites and alloys

Materials	Modulus (GPa)
ZCM630 SC	45
ZCM630+15%SiC SC	60
ZCM711 DC	46
ZCM711+12%SiC DC	55
ZCM711 Extruded	46
ZCM711+12%SiC Extruded	57

3.7. Coefficient of Thermal Expansion

Figure 13 shows the thermal expansion of matrix, SiC and composites. SiC, having a much lower thermal expansion than magnesium alloys reduced the overall expansion behaviour of composites. If the interface bonding between matrix and reinforcement is adequate and no debonding takes place during the thermal cycling, it has often been observed that the thermal expansion of composites lies between that predicted by a Rule of Mixtures and Turner's prediction which is based on Rule of Mixtures but also takes account of Poisson's ratio and bulk modulus of both matrix and reinforcement. Therefore it is more realistic approach than the Rule of Mixtures.

In this study the decrease in expansion coefficient of composites relative to the unreinforced matrices were found to be 12% for ZCM630+15%SiC SC composites respectively. These results seem to be close, but slightly less than expected from a Rule of Mixtures, although during the thermal cycling at temperature range of 20-350 °C, no debonding was experienced. By comparison to SC composites, the reduction in coefficient thermal expansion of ZCM711 DC and of extruded ZCM711+12%SiC composites were 23% and 32% respectively. These value are close to the lower limit (Turner prediction). Although the CTE of matrix alloys the same and also the SiC particle content of ZCM630+15%SiC SC composites is 3% higher than those DC and extruded composites, the former showed less reduction in the CTE than the latter. It is expected that with increasing volume fraction of reinforcement, there should be a significant decreases in CTE of composites.

The results indicate that microstructural defects particularly pores and hot tearing cracks, result in a small decreases in CTE for SC composites and result in an expansion that lies close to upper limit (Rule of Mixtures). However, with relatively fewer particles content ZCM711+12%SiC DC and more importantly extruded composite, demonstrate a remarkable reduction in CTE value that lies close to the lower limit. Therefore, (i) microstuctural defects, and (ii) interface debonding,(which was not observed any of these composites) allows the matrix a free expansion which eliminates the influence of reinforcement on the reduction of expansion. In this case the CTE of the composites will be close to that of the matrix material.

4. Conclusion

The introduction of SiC particles into magnesium base composites results in a marked change in microstructure;
Instead of a coarse grain and cell structure, a fine equiaxed grain structure was produced. During solidification, particles were mostly rejected by the growing magnesium grains and trapped into the grain boundaries where porosity was often observed. However, most of the grains contained a few intragranular particles.

Although there is no direct evidence that magnesium grains nucleated from the SiC particles, but particles could accelerate the nucleation of magnesium grains altering the heat diffusion and cooling rate during solidification. The eutectic phase $Mg(Cu\ Zn)_2$ however, nucleated at the SiC particles.

Agglomeration of particles due to inadequate consolidation and difficulty in wetting of particles with liquid magnesium was observed.

With 10% SiC particles in SC composites, the particles settling resulted in a 5mm particles free region at the top of the cast plate. In the DC composites with 12% SiC particles, a 2mm particle free region at the edge of the billet was also observed, but the particles free region in this case resulted from rapid solidification. This indicated that as much as slow cooling rate, very fast nucleation and cooling rate caused the particles rejection resulting a particles free region.

In the DC casting with relatively high solidification rate and improved composite consolidation fewer casting defects and a better particles distribution has been produced. After the extrusion, even more homogeneous distribution of particles with the elimination of most of agglomerated and clustered particles resulting in a pore free composite were obtained,

TEM study revealed that large thermal expansion mismatch between matrix and particles produced extensive twinning, a high dislocation density and a stress fields around the particles in all composites. Such dislocations and stress fields altered the precipitation kinetics of composites and produced a fine and dense precipitation around the particles and also resulted in more rapid ageing than that of unreinforced matrix during the heat treatment.

Internal stresses which are believed much higher in particles dense regions and around the particles than the average stress in the bulk composites developed in the cast composites. As a result of such stresses static recrystallization took place during an identical recrystallization heat treatment in the alloy (ZCM630+15% and ZCM711+12%SiC) composites producing a much finer grain size.

The addition of SiC particles to the magnesium matrices resulted in an increase in the elastic moduli and tensile strength of all composites. It was also found that good particle/eutectic bonding gave rise to an additional effect on the increase in elastic moduli. However, the casting defects produced a decrease in the moduli and elongation.

CTE expansion of the composites has also been influenced by the microstructural defects. Reduction in CTE of extruded ZCM711+12% SiC composite was 32%, whereas reduction in CTE of SC ZCM630+15% SiC composites with even high volume fraction of particle was only 12%, as a consequent of pores, hot tearing and oxide.

The hardness of composites increased between 15% and 40%. The improvement in hardness resulted from (a) the presence of SiC particles, (b) extremely grain refinement of particles
       (i) producing dynamic recrystallization during the extrusion,
       (ii) leading a static recrystallization in cast composites,
       (iii) increasing the nucleation side, blocking the grain to further growth during the solidification
and   (iv) causing a fine and dense precipitate in the matrix.

5. Acknowledgement

The author would like to thank Dr.T.Wilks, of MEL for supply of the materials and Mr.P.J.Harrington and Mr.A.Nichells for their assistance in preparing samples for this study. The author is also thankful to Dr.C.Hammond for his comments and suggestions.

6. References

1. B.A.Mikucki, S.O.Shook, W.E.Mercer and, W.G.Green, Die Cast Engineering 30,5, (1986), 26
2. B.A.Mikucki, W.E.Mercer and, W.G.Green, SAE Technical Paper Series 900533, International Congress and Exposition, Detroit, Michigan, February 26-March 2 (1990)
3. N.L.Hansen,T.A.Engh, O.Lohne, Interface in Metal-Ceramics Composites, Ed. by R.L.Yin, R.J.Arsenault, G.P.Martins and S.G.Fishman, TMMS, (1989)
4. E.M.Kiler, A.Mortensen, J.A.Cornie and M.C.Flemings, J. Mat. Sci,.26, (1991), 2519
5. B.Inem, The Development of Structure and Properties of Magnesium Matrix SiC Particles Reinforced Composites, Ph.D. Thesis, School of Materials, Metallurgy Department, University of Leeds, 1993.
6. M.Gallerneault and R.W.Smith, Cast Metals, 4, 3, (1991), 122
7. B.Inem and G.Pollard, The Second European Conference on Advanced Materials and Processes, "EUROMAT 91", Cambridge, Ed. by B.Clyne, The Ins. of Metals, vol.2, (1992), 127
8. B.Inem, J. of Mat. Sci., 28,1993, 4427
9. Y.L.Liu, N.Hansen and D.J.Jensen, Metall. Trans. 20A (1989)1743
10. F.J.Humphreys, W.S.Miller, M.R.Djazeb, Mater. Sci.and Tech. 6, 11, (1990), 1157
11. B.Inem, Mat. Sci. and Eng., (accepted for publication, 1995)
12. P.K.Chaudhury and H.J.Rack, J. Mater. Scie. 26 (1991) 2893

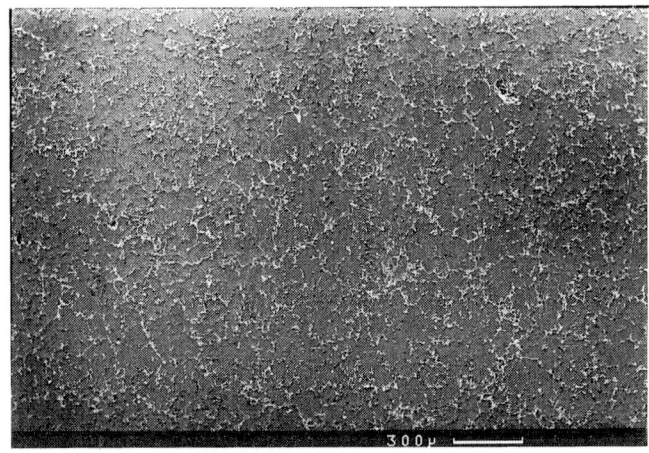

Figure 1
Micrograph showing distribution of the particles in
ZCM 630 composite with 15% SiC

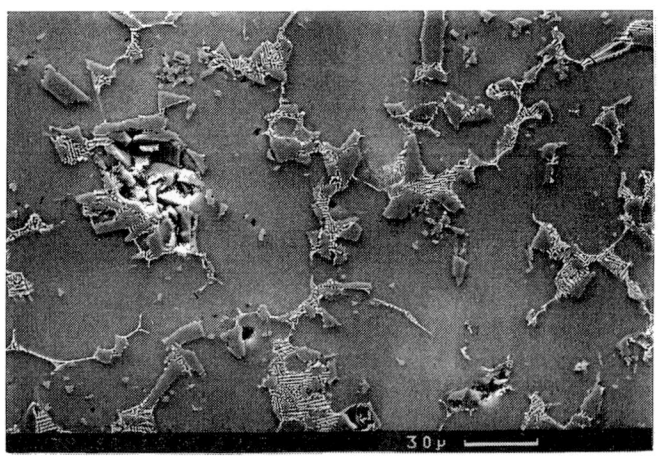

Figure 2
SEM micrograph showing particles segregation at grain
boundaries

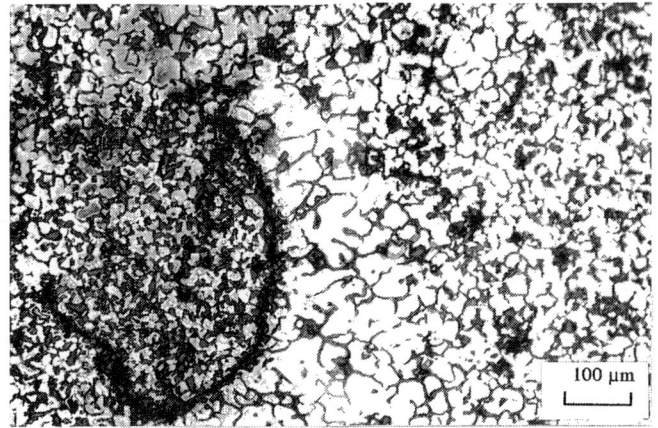

Figure 3
Optical micrograph showing clustering of SiC particles associated with oxide

Figure 4
SEM micrograph showing hot tearing in the matrix

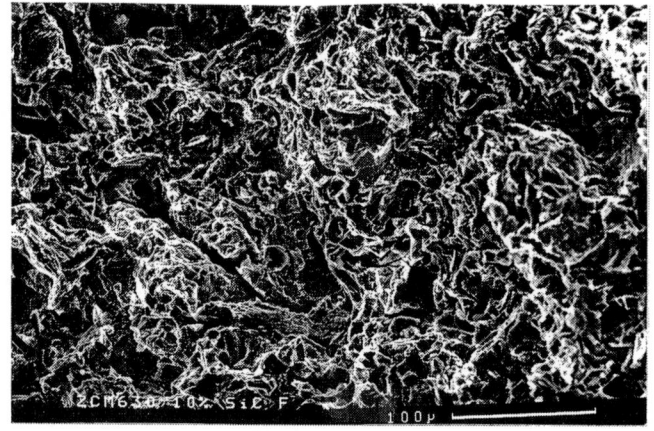

Figure 5
SEM micrograph showing intergranular cracks on the fracture surface

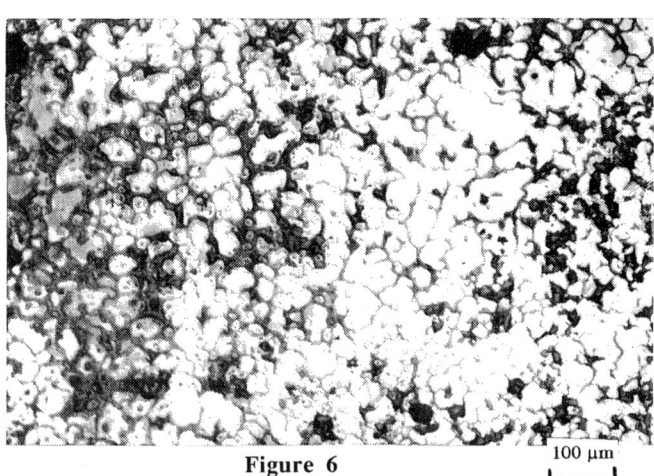

Figure 6
Optical micrograph showing particle free region at the outer part of DC billet

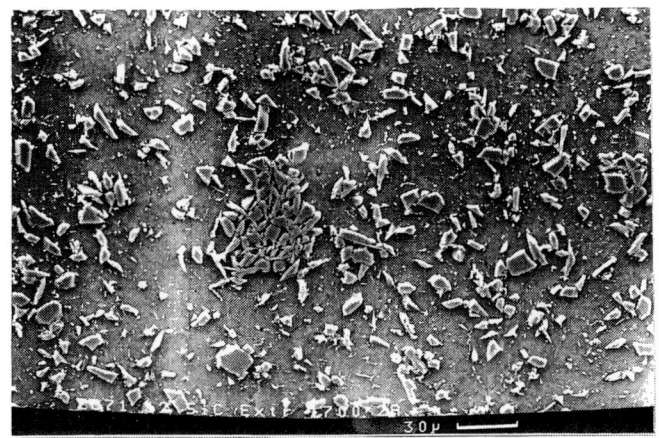

Figure 7
SEM micrograph showing the distribution of the SiC particles after extrusion

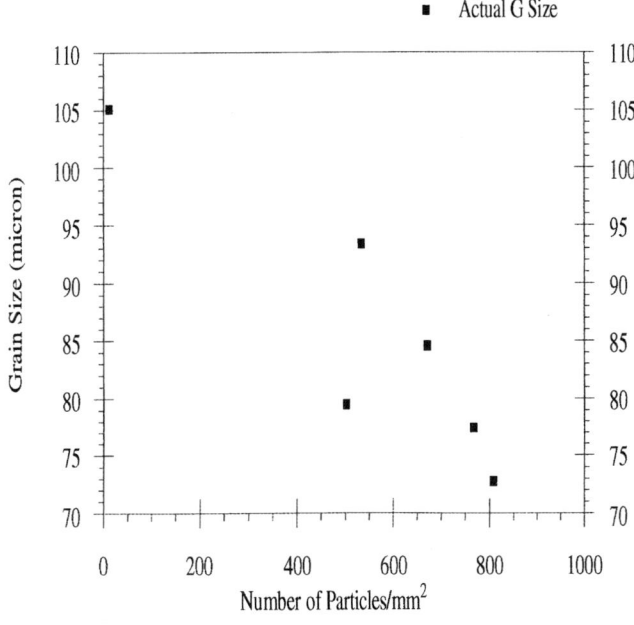

Figure 8
The relation between number of particles and grain size in

Figure 9
Optical micrograph showing static recrystallization in the CPMg composite with 15%SiC particles

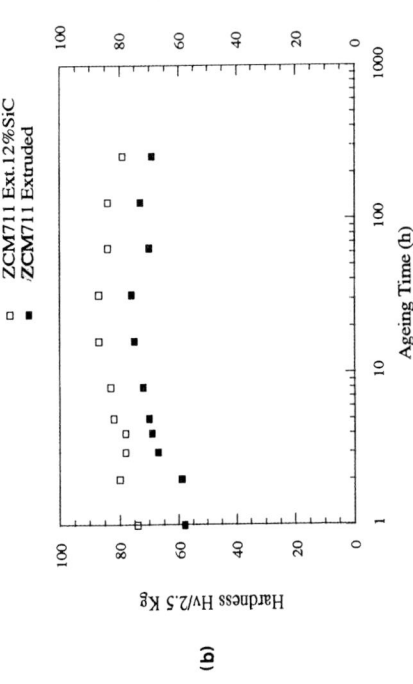
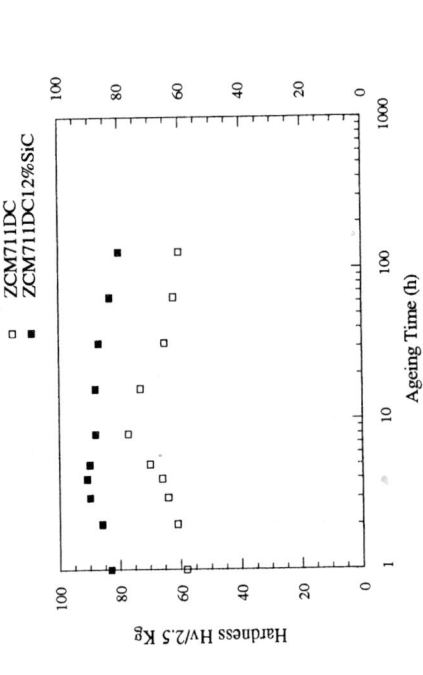
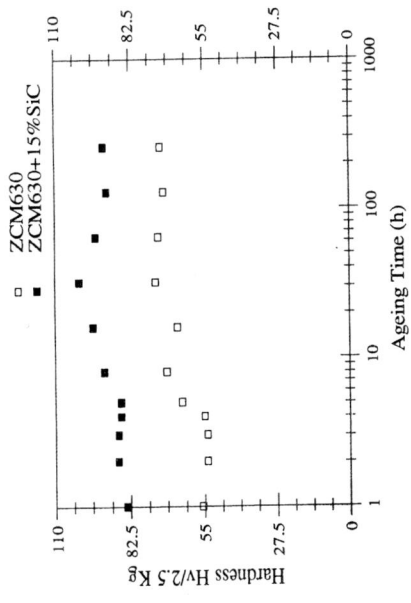

Figure 10

Ageing response of unreinforced
a. ZCM630 alloy and ZCM630 composite
b. ZCM711 Ext. alloy and ZCM711 Ext. composite
c. ZCM711 DC alloy and ZCM711 DC composite

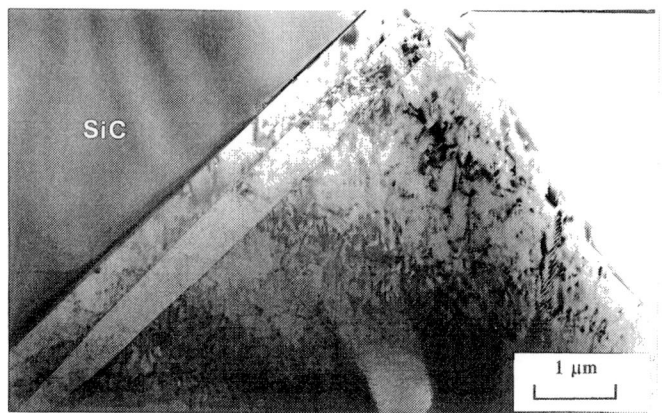

Figure 11
A high dislocation density and heavy twinning around the particles

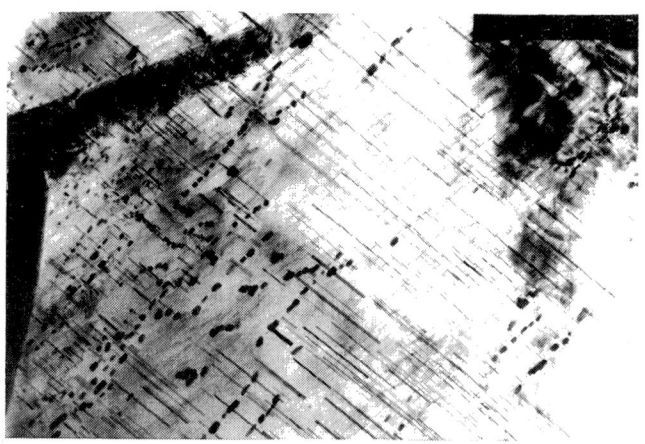

Figure 12
Dense precipitation in the regions near to SiC particles

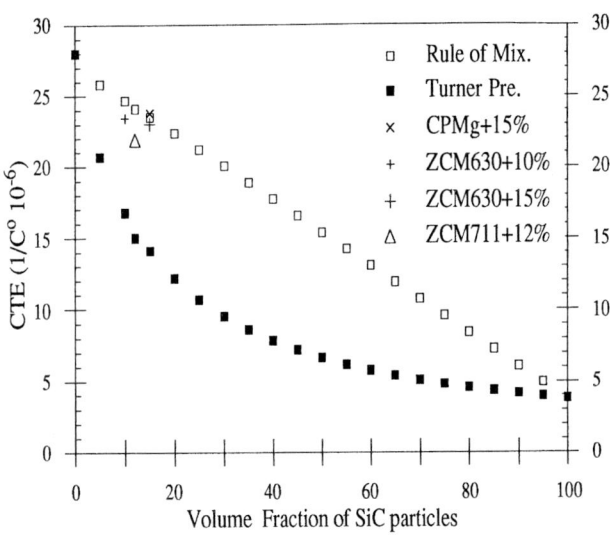

Figure 13
Predicted and measured coefficient thermal expansion of composites

FEASIBILITY OF ULTRASONIC INFILTRATION IN PREPARATION OF METAL MATRIX COMPOSITES

J. PAN, D. M. YANG AND H. WAN

Department of Materials Engineering and Applied Chemistry, Changsha Institute of Technology, Changsha 410073, P.R.China

Abstract *As a new application of high-intensity ultrasound, also a simple process to prepare metal matrix composites, the progresses of ultrasonic infiltration have been presented. The research results show that by ultrasonic infiltration process, fiber reinforced metal composite wire with a high strength can be fabricated, molten aluminium can permeate into a particulate preform, and pariculates can be dispersed in aluminium melt.*

Keywords High-intensity ultrasound, cavitation, fiber, particulate, light metal

1. INTRODUCTION

High-intensity ultrasound means an ultrasonic vibration with a high power. Recently, its application has been developed in material preparation industry, such as ultrasonic treatment of light alloys on the processes of nucleation and growth of crystals during solidification of ingots and casting[1], ultrasonic vibration compaction of powder and then preparing fiber reinforced aluminium alloy composites by a metallugical technique [2,3], ultrasonic welding for different metals[4,5]. In metal matrix composites, there are two main problems that are unfavourable for fabricating composites and enhancing their mechanical properties. One is the wettability of fiber by liquid aluminium. For example, previous research showed that the wetting angle θ between SiC(Nicalon)fiber and aluminium is<90° only when the temperature is above 1050°C; so the wettability between them is not good[6]. Another is the detrimental interfacial reaction to degrade the fiber strength seriously, which means that the compatibility between fiber and metal is poor[7]. With the general fabrication process, it is difficult to overcome these two problems. An ultrasonic infiltration method to prepare SiC yarn fiber reinforced aluminium composites, was developed successfully in 1985[8,9]. By unceasing improvements in experiment apparatus and processing parameters, now it is possible to prepare SiC/Al composite wire with a continuous length longer than 500 meters and an ultimate tensile strength higher than 1500MPa[10]. As further application of this ultrasonic technique, SiC/Al composite tape, CF/Al composite wire, SiC(CVD)/Al composite wire and the woven fiber cloth reinforced aluminium plate have also been developed. Few papers were found that focused attention on the ultrasonic infiltration technique of preparing SiCp/Al[11], SiCw/Al[12] and Al_2O_3/Al[13, 14] composites. We also conducted a wide research on this technique for fiber, whisker and particulate reinforced aluminium or magnesium composites[15]. This paper aims at a review on some research progresses of ultrasonic infiltration for metal matrix composites preparation process and the infiltration mechanism explanation.

2. EXPERIMENT METHODS

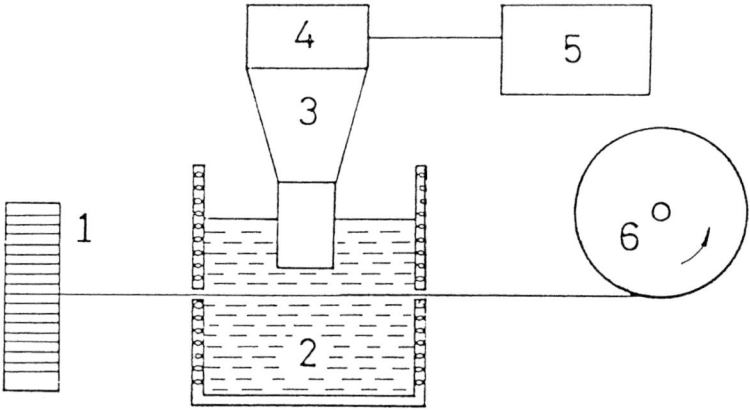

Figure 1 A schematic diagram of fabricating continuous fiberreinforced metal composites wire: (1) fiber, (2) crucible containing metal melt, (3) ultrasonic horn, (4) transducer (5) generator, (6) composite wire winding wheel

The apparatus for fabrication of a continuous fiber reinforced aluminium or magnesium composites by the ultrasonic infiltration method is shown in Fig. 1. The fiber was pulled continuously through a crucible(2) which containing molten metal. An ultrasonic energy was imputted into the melt via an ultrasonic horn(3) attached to the transducer(4). The fiber cannot be wetted by metal melt without the ultrasonic vibration. Under the action of ultrasound the fiber was infiltrated into the molten metal and became a composite wire after passing through the crucible simultaneously. For discontinuous fiber reinforced aluminium, a preform of particulate or whisker was placed below the ultrasonic horn and hence infiltrated by or dispersed into aluminium melt.

As continuous reinforcement, we used a multifilament SiC(Nicalon) fiber, which is manufactured by Nippon Carbon Co. Ltd., with a diameter of 12μm and 500 filaments for a yarn. The laminated preform by the woven SiC(Nicalon) fiber cloth was also used to prepare a composite plate. Another SiC multifilament(Tyranno) with a diameter of 11μm and 800 filaments for a yarn, was the product of UBE Industries, Ltd. Carbon(T300) fiber, fabricated by Toray Industries, Inc., with a diameter of 7μm and 6000 filaments was used. A CVD-SiC monofilament(SCS-2) supplied by Textron Specialty Materials Inc., having a diameter of 142 μm, was used too. As the discontinuous reinforcement, a commercially available SiC particulate (d=100μm) and a TiO_2 whisker supplied by Titan Kogyo Co. Ltd. (d=0.2~0.8μm, l=13μm) were used. An industrially pure aluminium(Al>99.6wt% purity), Al-Mg alloy(Mg 5.8~6.8wt%) and a magnesium alloy(Al 7.5~9.0, Mn 0.15~0.5, Zn 0.2~0.8wt%, Mg bal.) were used as the matrix.

The ultimate tensile strength of composite wire was tested under the conditions of gauge length 50mm, cross speed 0.5mm/min for 10~20 samples as an average value. TEM observation on the interface of SiC(Nicalon)/Al was carried out by a HITACHI-H800 transmission electron microscope.

3. MECHANISMS OF ULTRASONIC INFILTRATION

It is commonly known that ultrasound can be divided into two types, called detective ultrasound and high-intensity ultrasound, respectively. The former is used for gathering messages by means of its penetrability through material such as ultrasonic detection of defects and medical ultrasonic diagnosis, the latter is for treating material by using its high energy such as ultrasonic cleaning, ultrasonic weld and ultrasonic machining. High-intensity ultrasound has a general frequency in the range of 10^3~10^4Hz. The ultrasonic generator power used in the present composites preparation was from 0.5~2kw. High-intensity ultrasound theory states[16] that when an ultrasonic energy is introduced into a liquid, a temporary negative pressure arises in some liquid regions owing to the vortex flow or other physical effects. An excessive tensile stress induced creates bubbles or voids in the regions. Those bubbles are in an unsteady state and will oscillate under the action of ultrasonic field. If the negative acoustic pressure exceeds some certain value, the bubbles expand 100~1000 fold rapidly and subsequently collapse or implode at a fast speed. While the bubbles collapse, a shock wave generates a powerful hydraulic pulse of 500~1000MPa or creates cumulative jets with a speed about 50~100 m/s[1]. This behaviour is called the cavitation effect of high-intensity ultrasound. The bubbles that give rise to the cavitation effect are called cavitation nucleus. In the case of the equipment used here, after an ultrasonic horn was inserted in a beaker containing 500ml water, the generator frequency and power were adjusted till the ultrasonic transducer was in the resonant state. At that moment, it was readily seen that there was a tail flow below the horn end and the liquid was full of many movingbubbles. If some particulates were put into the beaker, they would be dispersed in water under the action of ultrasonic vibration(see Fig. 2). When the ultrasonic horn was taken off the water level, the water that stayed on the horn end atomized rapidly, forming a water jet as shown in Fig. 3, which also demonstrated that the ultrasonic transducer was in a resonant state.

Figure 2 A dispersion system showing particulates into water by the action of ultrasound: (a) power is off, (b) ultrasonic transducer is in the resonant state.

Figure 3 An atomized water jet from the ultrasonic horn end in resonant state.

When we imput ultrasound into a molten metal melt(at about 700℃) and adjust the generator frequency and power, until the transducer is in resonant state, the fiber that is not wetted by metal melt originally will be wetted and consequently be infiltrated into by the metal melt, and hence become composites. It can be considered that the ultrasonic cavitation is a reason for ultrasonic infiltration. The air brought in metal melt by fiber and the gas dissolved in the melt may become the cavitation nuclei. If the nuclei collapse, a high pressure or a cumulative jets created around the individual fiber, will be easy to carry the melt into the fiber. Because the collapse speed is extremly high, the fiber and the metal melt combined to be composites very rapidly. In addition, the ultrasonic thermal effect forms a high temperature field near the fiber, which is favourable for improving the wettability of the fiber by liquid metal.

4.PREPARATION OF FIBER/METAL COMPOSITES

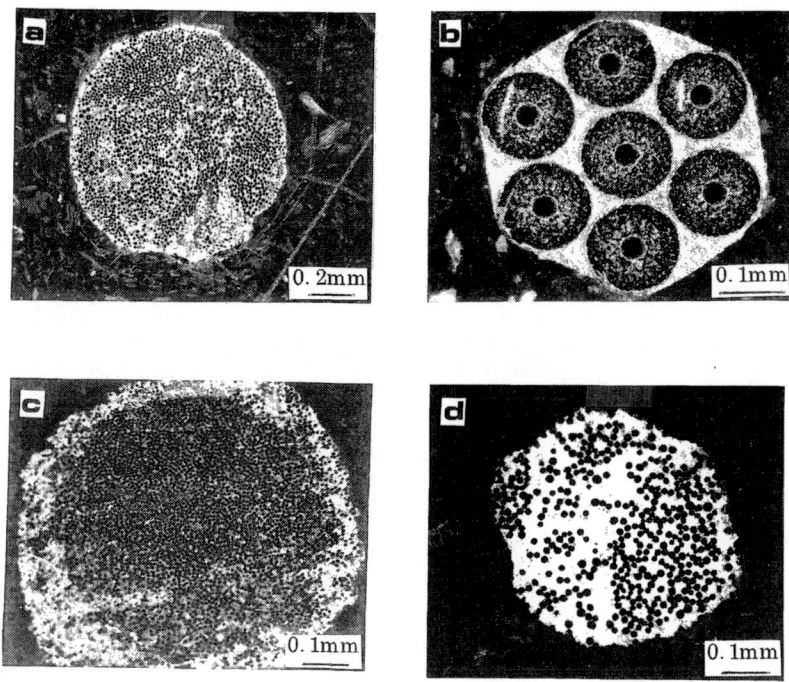

Figure 4 Optical micrographs of composite wire section: (a)SiC(Nicalon)/Al, (b)SiC(SCS-2) /Al, (c) CF(T300)/Al, (d) SiC(Nicalon)/Mg.

After the fiber passes through the ultrasonic vibration system shown in Fig.1, it is fabricated to a continuous composite wire. Fig.4 shows the metallographic photographs of different kinds of fiber reinforced aluminium or magnesium composite wire. It can be seen that the fiber distributes in metal matrix homogeneously, without obvious casting defects. Table 1 lists the properties of composite wire. In the fabricated processing of composite wire, the contacting time between the fiber and metal melt can be controlledvery short, for instance, in the case of SiC(Nicalon)/Al which is about one second. Hence the

Table 1 Properites of fiber/metal composite wire

wire	σ_f (GPa)	ϕ (mm)	V_f (%)	σ_{wire} (MPa)	σ_{ROM} (MPa)
SiC(Nicalon)/Al	2.5	0.45	48	1543	1255
SiC(Tyranno)/Al	2.5	0.45	45	1530	1180
SiC(SCS-2)/Al	4.6	0.42	84	1928	3880
CF(T300)/Al	3.0	0.50	47	513	1463
SiC(Nicalon)/Mg	2.5	0.55	30	765	855

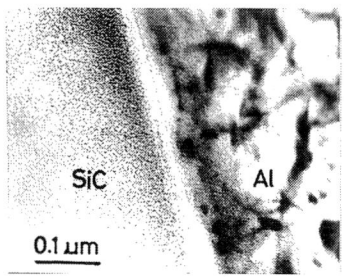

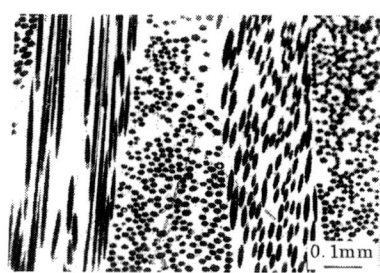

Figure 5 TEM micrograph showing a clean interface between SiC(Nicalon) fiber and pure aluminium matrix.

Figure 6 Optical micrograph of SiC(Nicalon) fiber woven cloth infiltrated by aluminium.

detrimental interfacial reaction is little, a higher strength wire can be prepared. Fig. 5 shows a TEM micrograph of SiC(Nicalon) -Al interface which is fairy clean, no any reaction product appears at the interface. The extracted fiber strength maintained a higher level, comparing with the original one. The wire strength was corresponding with the predication value by rule of mixture(ROM). Another SiC(Tyranno) yarn fiber reinforced aluminium wire also had a high strength. But for SiC(SCS-2)/Al, CF(T300)/Al and SiC (Nicalon)/Mg composite wire, there was still an improvement in fabrication process because the wire strength had an obvious degradation, comparing with their ROM value.

In the case of SiC(Nicalon) woven cloth, while the cloth was placed below the ultrasonic horn, the general horn suited to prepare a composite wire could not be efficient for the cloth infiltration. Only by using an ultrasonic horn with a higher energy density, prolonging the period of ultrasonic infiltration on fiber appropriately and adjusting the other processing parameters could we fabricate SiC(Nicalon) fiber cloth reinforced aluminium composites successfully. The optical microstructure of this composite plate is shown in Fig. 6. The thickest composite plate we prepared was one that contained a 7-ply yarn fiber. In further research we are considering how to fabricate a big composite plate or a continuous composite tape. Furthermore, if a higher power ultrasonic equipment is applied, it would be possible to realize the infiltration of aluminium into a preform that is knitted with the yarn fiber to prepare a near net shape mechanical component by ultrasonic liquid infiltrating directly.

5.PREPARATION OF PARTICULATE/METAL COMPOSITES

According to the behaviour of high-intensity ultrasound, the ultrasound has an ability of improving the wettability and dispersing solid particulate into liquid. By using those features, SiC particulate reinforced aluminium preparing system was designed. If an ultrasonic vibration acted on SiC particulate preform which inserting in aluminium melt, a SiCp/Al composite plate was obtained(Fig. 7a). The infiltration deepness of aluminium depended upon the ultrasonic energy density. The deepest one was 6mm for the present apparatus. If SiC particulate was poured into aluminium melt vibrated ultrasonically, they could be dispersed in aluminium(Fig. 7b). But in this case the volume

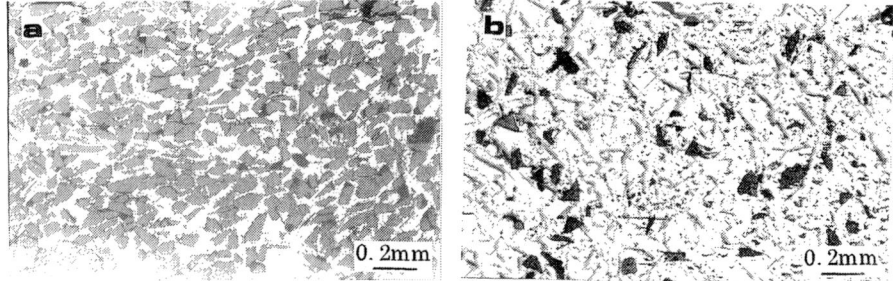

Figure 7 Optical micrographs of SiC particulate reinforced aluminium composites:
(a) Al infiltrated in SiCp preform, (b) SiCp dispersed into Al melt.

fraction of SiC particulate was lower($<10\%$).

In order to obtain an oxide particulate reinforced aluminium composites by in-situ process firstly, the differential thermal analysis(DTA) was carried out. From the result, the following deoxidation reaction is thought to occour at the temperature range of 710~770℃:

$$Al + TiO_2 \rightarrow Al_2O_3 + (TiAl, Ti_3Al, TiAl_3) \quad (1)$$

In the process of reaction squeeze casting or reaction powder metallurgy, Al and TiO_2 may not react perfectly and the structures of composites may not be homogeneous. In the present work, a TiO_2 whisker was added to the Al-Mg alloy melt which was subjected to the ultrasonic vibration system. The result shows that TiO_2 whisker preform could be broken and dispersed in the melt and then reacted with Al-Mg alloy by the following reaction:

$$(Al-Mg) + TiO_2 \rightarrow MgAl_2O_4 + TiAl_3 \quad (2)$$

In other words, an oxide particulate $MgAl_2O_4$ and an intermetallic compound $TiAl_3$ reinforced Al alloy composites were prepared. In Fig. 8 the two kinds of reinforced phase

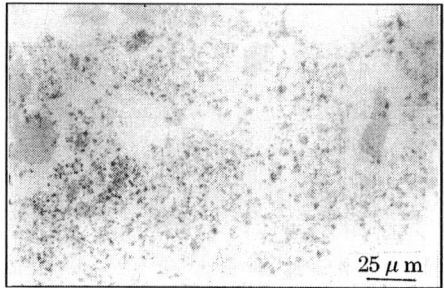

Figure 8 Optical micrograph of $MgAl_2O_4$, $TiAl_3$ reinforced Al-Mg alloy composites.

can be observed, one is a fine particulate with the size less than $1\mu m$ and another is a larger gray phase. X-ray diffraction analysis shows that the secondary phase consists of $TiAl_3$, $MgAl_2O_4$ and MgO. The further research is how to control the reinforcement size and its distribution.

6. SUMMARY

From a number of composites system investigated, it is considered that the high-intensity ultrasound is an effective approach to solve the problems containning wettability and compatibility when fabricating a continuous or discontinuous fiber reinforced metal, The main mechanism of ultrasonic infiltration is thought to be the ultrasonic cavitaion which creates a powerful pressure and a high temperature field around the reinforcement, then permeates the molten metal melt into fiber bundle or particulate preform. By the same effect, particulate can by dispersed in metal melt. Because ultrasonic infiltration may be completed rapidly, the fiber damage resulted from the contact of fiber with liquid metal can be minimized.

ACKNOWLEDGMENTS

The authors are grateful to J. W. Li for preparing SiC cloth reinforced aluminium and to J. S. Hu for help with the TEM observation of SiC/Al composite wire.

REFERENCES

[1] G. I. Eskin, Ultrasonic Sonochemisty, 1, S59(1994)
[2] J. Tsujino, Ultrasonic International'89 Conference Proceedings, 346(1989)
[3] S. Wakayama and H. Nishimura, J. Japan Inst. Light Metals, 38, 672(1988)
[4] J. Tsujino, H. Furuya and Y. Murayama, Ultrasonic International'89 Conference Proceedings, 268(1989)
[5] S. Matsuoka, New Ceramics, 8, 67(1993)
[6] D. M. Yang, A. M. Chen and Z. M. Zhang, ISCMS Abstracts of Papers for Work-in-Progress, edited by T. T. Loo and C. T. Sun, Technomic Publishing Co., Inc., Pennsylvania, 45(1986)
[7] J. Pan and K. T. Wei, Aerospace Material & Technology, 1, 22(1985)
[8] D. T. Huang, J. Tan, Y. K. Hao and J. Pan, Proc. of '85 Conference on MetalMatrixand Ceramic Matrix Composites, Hangzhou, China, 15(1985)
[9] D. T. Huang, D. M. Yang, X. F. Yin and J. Pan, Materials Engneering, 3, 17(1989)
[10] D. M. Yang, X. F. Yin, and J. Pan, Mater. Sci. Lett., 12, 252(1993)
[11] R. F. Orban and D. M. Goddard, AD-A183960(1987)
[12] C. J. Skowronek, A. Pattnaik and R. K. Everett, AD-A168836(1986).
[13] H. Nakanishi, Y. Tsunekawa, M. Okumiya and N. Mohri, Materials Transactions, JIM, 34, 62 (1993).
[14] H. Nakanishi, Y. Tsunekawa, N. Mohri, M. Okumiya and I. Niimi, J. Japan Inst. Light Metals, 43, 14(1993)
[15] J. Pan, C. Li, D. M. Yang and X. F. Yin, Metal Matrix Composites(ICCM/9 Proc.), edited by A. Miravete, Uni. of Zaragaza Woodhead Publishing Ltd., 1, 801(1993)
[16] M. G. Sirotyuk, High-Intensity Ultrasonic Fields, Part V, edited by L. D. Rozenberg, Plenum Press, New York-London, 263(1971)

CASTING OF COMPOSITE COMPONENTS

S. Ray
Department of Metallurgical Engineering
University of Roorkee
Roorkee (U.P)-247667,
INDIA

Abstract

The article describes the routes to stircasting or compocasting of composite components for engineering applications. The problems associated with processing of slurry of molten metal containing dispersoid particles, its transport to casting bay and casting of slurries in moulds, have been discussed. The possible causes of particle segregation in cast composite microstructure have been outlined in the context of investigations carried out so far.

1. INRODUCTION

The primary reason for the development of Metal Matrix Composites (MMC) so far, is its ability to offer a spectrum of properties tailor made for a given application. Substitution of existing components by one made of composite materials, has a considerable potential for weight saving and this prospect generated interest in defence, space and aerospace industries who are always looking for light and reliable materials because of high premium associated with weight saving in these industries. However, the interest of these industries could not generate widespread R&D activities in this area as they are low volume user of materials. In the eighties, the automobile industries have taken considerable interest in composites because of the potential for weight saving and durability of composite components in comparison to their alloy counterparts. Although the automobile industries are high volume user of materials, the economic premium for weight saving in a automobile is about 1/30th of that in an aircraft or 1/300th of that in missile[1,2]. Therefore, the interest of automobile industries are primarily in cheaper composites resulting from solidification processing. Stircasting or compocasting in which dispersoids are stirred in a liquid or semi-solid alloys and the resulting slurry is cast into components, offers a unique advantage in terms of capital investments. These methods can be easily adapted by a foundry with its existing infrastructure and therefore, are relatively more attractive to the industries in comparison to other solidification processing routes. The present article describes the different steps involved in stircasting or compocasting, the necessary consideration involved in each step and the state of knowledge and the existing gaps where further investigations should be directed.

2. MELT-DISPERSOID SLURRY

The disperesoids commonly used to prepare slurry are either particles, short fibers or whiskers having a typical dimension of 10 to 100 μm. The dispersoids of commercial interest are silicon carbide, alumina, carbon, graphite, TiB_2, Nb_2Si or $MoSi_2$ etc. The melt in the slurry can result from a metal, an alloy or an intermetallic compound. Light weight alloys based on aluminium and magnesium have engaged the attention so far because of a stress on weight saving. But special wear resistant composites based on Iron base alloys have been developed by stircasting to find cheaper substitutes for commercial PM `Ferro-TiC' Materials[3]. Composites based on aluminides have been developed for high temperature applications[4]. A variety of base alloys both ferrous and non-ferrous, will be explored in future as and when composites will be developed for different new applications.

Once the matrix alloy and the dispersoids have been chosen for a given application, slurry has to be prepared by melting the matrix alloy and stirring the dispersoids in it. Since the dispersoids have very small dimension, the entry of these dispersoids into the molten alloy will depend on the surface forces[5,6] which have higher magnitudes compared to the forces of gravity or buoyancy. It is easier to disperse the dispersoids into the melt if the melt is able to wet the dispersoid. Generally, the dispersoids of commercial interest like SiC or Al_2O_3 are not wetted by molten aluminium as shown in Table-I, from the contact angles determined by ssesile drop experiments reported in literature[7,8].

There are several approaches for promoting wetting of dispersoids by melt. In the pioneering investigation on solidification processing, Badia and Rohatgi[9] coated graphite particles by nickel in order to make them wettable by molten aluminium. The author in his attempt to develop $Al-Al_2O_3$ particle reinforced composite in 1969, used alloying addition of magnesium in molten aluminium for the promotion of wetting of alumina by molten aluminium[9]. At lower level of addition of magnesium, it was found that magnesium reacts with alumina particles on the surface and forms a layer of $MgAl_2O_4$ which is wettable by molten aluminium[9]. Treatment of dispersoids by inorganic compounds like K_2ZrF_6 has improved wettability of SiC particles by molten aluminium[10] presumably because of removal of surface oxide layer over molten aluminium. Promotion of wetting by coating or reactive alloying or inorganic solution treatment should be practiced with caution. There should not be any adverse impact on the properties of composites as a consequence. Since reactive wetting often results in a product(P), the M/D interface in the resulting composites, is substituted by M/P and P/D interfaces. If M/P or P/D interfaces have lower fracture strengths than M/D interface, the strength of a structural composite may be adversely affected[11,12]. If one uses coating which is soluble in the matrix alloy, the thickness of such coating on dispersoids should be such that it remains there during processing.

2.1 PROCESSING CONSIDERATIONS

Once the matrix alloy and the dispersoids have been chosen on the basis of the spectrum of properties required for a given application and a suitable strategy has been charted to promote wetting, one may start the processing to prepare melt-particle slurry for stircasting or compocasting. The important considerations for this step are (a) processing temperature, and (b) the details of stiring.

In monolithic alloys the pouring temperature of an alloy for casting is kept high enough above liquidas temperature so as to possess the requisit casting fluidity. But at the same time, the temperature should not be too high to result in high solubility of gases and the consequent increase in gas porosity during casting.

But in melt-dispersoid slurries, the shear stress is not linearly proportional to shear rate and the ratio of shear stress to shear rate, termed apparent viscosity, η_a, reduces as the shear rate increases. Thus, the slurries show shear thinning behaviour because the cluster of dispersoids in the slurry are broken at higher shear rates reducing the resistance to flow and η_a may be expressed as,

$$\eta_a = \tau/r = k(r)^m \qquad (2)$$

Worsten et al[15] have determined the value of m as - 1.3 for a slurry of A356 aluminium alloy containing 18 vol% of 13μm SiC particles. Moon[16] has determined apparent viscosities for different shear rates in a number of Al-alloy-SiC slurries and .2 shows some of the results obtained by him.

The apparent viscosities of slurries containing about 30 vol% of Silicon carbide particles are observed not to exceed 100 cp even at the beginning of stirring. During dispersion of Silicon Carbide in Semi-Solid alloys, a slurry containing 40 volume fraction of solid constituents of primary solid and dispersoid particles, show an apparent viscosity lower than 100 CP. Therefore, under these circumstances a turbine stirrer should suffice. There are two types of turbine stirrers radial and axial, classified on the basis of direction of discharge. In flat blade radial turbine stirrers the discharge is perpendicular to the centerline of the crucible but in pitched blade axial stirrers, a component of discharge parallel to the centerline of the crucible is also created.

The axial flow parallel to the centerline of the crucible helps in lifting particles denser than the liquid from the bottom and counteracts its settling tendencies. The state of dispersion has been found to correspond well with a composite parameter called Particle Dispersion Number (PDN) which has in it, the ratio of velocity of axial flow to the terminal selting velocity and it is given as,

$$PDN = H_0(\eta\Omega)^{1/2}/[r_{i１/4}d^{3/4}V t] \qquad \ldots(3)$$

where, H_0 is the height of the melt, Ω is the angular velocity of the Stirrer of radius r_i and d is the gap between the stirrer and the crucible wall. η is the viscosity of the slurry and v_t is the particle settling velocity. El-Kaddah and chang has recommended that PDN should be greater than four for homogeneous dispersion[17]. The speed of the stirrer may be worked out on the basis of PDN under a given stirring condition.

It has generally been recommended that the turbine stirrer should be so placed as to have 35% liquid below and 65% liquid above it. If the depth of liquid below an axial stirrer is less than 30% of its diameter, the vertical discharge is throttled by the reflection of back pressure and the effectiveness of stirring is reduced. For such situations, radial turbine stirrer is a better choice. If for some reason, the stirrer is so placed as to exceed 65% liquid above or 35% liquid below it, multiple turbine stirrer

The same considerations are valid for pouring and processing temperature of a slurry in general but there are additional considerations. Higher processing temperature generally results in better wetting but in certain systems like SiC/Al-alloy there is a limiting temperature above which undesirable chemical reaction takes place at matrix-dispersoid (M/D) interface affecting the properties of composites adversely.

In compcasting, Processing of slurry is carried out at a temperature lower than the liquidus temperature because the particles of primary solid phase helps retention of dispersoid particles which undergoes random collision with primary phase particles in the slurry counteracting its settling or floatation governed by gravity and buoyancy[13,14]. In other words, the effective viscosity of the semi-solid alloy is higher than that of molten alloy and so, the stokes settling and hundered settling rates are reduced. However, the lowering of processing temperature below that of liquidus is associated with a penalty in terms of reduced casting fluidity and also an enhanced shrinkage porosity.

The dispersoids could be dispersed in molten alloy by several techniques. Badia and Rohatgi[9] used a gas injection system which used argon gas to carry powders of dispersoid into liquid alloy and disperse the particles during bubbling. The present author[7] used for the first time a stirrer to disperse particles and subsequently, this technique is followed widely for dispersion of particles in stircasting or compocasting method. The details of stirring includes stirrer design and dimension apart from the speed and position of stirrer in melt. It has been observed that the state of dispersion and the amount of dispersoids suspended in the slurry state are influenced by the details of stirring.

There are a large number of design of impellers which may be used for stirring low viscosity and high viscosity meltsas shown in . 1 Paddle, propeller and turbine stirrers are used for stirring low viscosity melts while anchor, helical ribbon and helical screw stirrers are generally used for stirring high viscosity melts. Propellers create axial flow while flat bladed turbines result in radial flow. If the blades of a turbine are angled, a combination of radial and axial flow may be created. Paddles are similar to turbines in flow characteristics. Turbine and propellers operate best under turbulent flow conditions at viscosities upto 20 poise for propellers and 600 poise for turbines.

A slurry of molten metal/alloy containing dispersoids has more resistance to flow than that in the corresponding melt. Molten metals and alloys generally behave as a Newtonian fluid and the shear stress, τ, required to initiate and maintain laminar flow is linearly proportional to the velocity gradient or shear strain rate r. The proportionality constant known as the coefficient of viscosity, η, characterises the resistance to flow and is written as,

$$\eta = \tau/r = \text{const.} \qquad (1)$$

should be used. Two turbines are generally necessary if the slurry depth exceeds 1.3 times the crucible diameter and the upper one should be placed at a depth between 0.5 to 1.0 times the crucible diameter, below the top surface of the slurry. For bottom pouring two turbines are better than one. The recommendations in this paragraph summarizes the experience of chemical engineers who generally deal with mixing of wettable particles. For poorly wetting melt-particle system, particle incorporation may also pose a problem because of surface energy barrier associated with its immersion into molten alloy. Under such circumstances stirring will have to help the particle to overcome this energy barrier. Therefore, these recommendations may have to be modified in the context of mixing of non-wettable melt-particle system.

Vortex created during stirring sucks even nonwetting particles into molten alloys but along with it may also suck bubbles. Often, non wettable particles attached to the bubbles, increases its weight counteracting buoyancy which helps it to float out. As a result, it has been observed that porosity content in a stircast composite varies almost linearly with particle content. as shown in 3 for composites synthesized over widely different stirring conditions and temperature. Ghosh and Ray[19, 20] have found that the particle incorporation is Maximum at a stirring speed of 16 revolution S^{-1} when a stirrer having diameter 63 percent of that of crucible, is placed in semi-solid alloy at 900K at a depth of 19 percent of total depth of liquid from the top. This relatively lower depth at which stirrer is placed as compared to that recommended by chemical engineers, has helped to expose a part of the stirrer and the surface of the liquid at the centre of vortex, descends below the stirrer. This condition has been found to enhance the particle incorporation significantly because of more effective particle transfer. Further experiments are necessary to optimize the design of stirrer and find out conditions which will not only achieve a homogeneity in slurry but at the same time, will take care of the requirements of particle transfer into a slurry of nonwetting particles.

Mixing time is one important variable which is often not adequately recognized or reported. Many of the metal-ceramic systems of commercial interest are made wettable by promoting interfacial reaction or segregation. Since these processess interfering with interfacial energy balance, progress with time, the contact angle, θ, is often a function of time as shown in Table-I. Therefore, if the processing time is short, the dispersoids may appear nonwetting but with an increase in time, these dispersoids become wettable. Similar time dependence of contact angle may also be observed for coated dispersoids if coating is soluable in the melt. The processing time should be controlled so that coating does not dissolve completely. Also, mixing to a steady state slurry with relatively uniform distribution of dispersoids, may take some time but we are not aware of its magnitude.

Once the processing conditions in terms of details of stirring and temperature, have been decided, one is in a position to start the process of mixing dispersoids into melt as it will be described in the following section.

2.2. PROCESSING OF DISPERSOID-MELT SLURRY

A schematic set-up[13] used by the author to mix dispersoids into alloy melt is shown in .4. A crucible placed inside a resistance heated or induction heated furnace, contains alloy heated to a temperature above its liquidus temperature. Wetting additions, if necessary, are made and the furnace is switched off to reduce the temperature to the processing temperature, above or below the liquidus depending on whether one is interested in stircasting or compocasting. If the processing temperature is below the liquidus and primary solid phase forms, it will be necessary to stir the melt while cooling to break dendrite morphology of primary phase[21]. Once the processing temperature is attained and the appropriate stirring conditions have been created, dispersoids are added to the melt and mixed by stirring. The rate of addition of dispersoids should be such that there is no accumulation of dispersoids at the top of the melt. After the dispersoids have been added, the melt is stirred for some more time for getting a homogeneous slurry which is then poured through the bottom of the crucible into a mould, by removing a plug while stirring continues. It is also possible to vary processing temperature such that particle incorporation and mixing is carried out at a temperature below the liquidus and then the slurry is heated up to above the liquidus temperature before pouring so that there is no loss of fluidity and a reduced shrinkage porosity.

The mould may be a permanent mould or sand mould. However, the mould design will have to be modified for casting slurry as recommended by the commercial manufacturers of composite ingot, described in Table-II. Depending on cost, quality and production volume, other special techniques of casting like die casting, may also be used to manufacture components.

The slurry for composite casting may not be prepared afresh as there are now commercial manufacturers of composite ingots from whom one may purchase ingots and remelt. But these ingots may have been cast under special conditions of vacuum or inert gas cover for reducing porosity. Unless special precautions are taken, the quality of composite components might not be as good as cast ingots. Table-II outlines a list of recommended precautions which should be taken during remelting.

It should be noted that during remelting, no fluxing or degassing is recommended. Many of these fluxing compounds containing halides are suspected to affect contact angle adversely causing rejection of particles from the slurry. Degassing by bubbling inert gas often results in poorly wetting particles clustering to these bubbles. As a result, the particles either float up or the bubbles stay behind inside the slurry resulting in porosity. However, inert gas cover has been recommended to check oxidation.

Difference in density between the dispersoids and the alloy melt results in settling of particles during remelting. Stirring is recommended to counteract settling. A special design of fluted stirrer is claimed to keep the top of the melt relatively quiet and therefore, may not result in high porosity due to enhanced dissolution of gases at the top. However, this expectation has not been fulfilled in a study conducted by a reputed foundry which recommends use of an open riser and a skimmer core to float out the gas bubbles and oxides[22].

2.3 SETTLING OF PARTICLES IN A SLURRY

Remelted or freshly prepared melt particle slurry will have to be transported to the casting bay for casting components in industrial foundries. One may arrange to stir a slurry even during transportion in a ladle, till it is cast. But a simpler approach will be to control slurry characteristics so that there is no significant degradation of slurry over a time period before which it is cast. A study of settling characteristics of a slurry is necessary in order to be able to control settling.

Stokes derived an expression for terminal settling velocity of a single rigid sphere through an unbounded quiescent Newtonian fluid at zero Reynolds number by equating viscous drag with net downward force of gravity as modified by buoyancy, as

$$V_s(r) = 2(p_p - p_l)gr^2/9\mu \qquad \ldots(4)$$

where r is the radius of the particle with density p_p. μ is the viscosity of the fluid with density p_l. g is the acceleration due to gravity.

This simple picture of stokes settling does not hold for slurries because of interaction between particles during settling. Richardson and Zaki[23] proposed an empirical relation for settling velocity of a particle, $V_h(r)$, in a monodisperse slurry when its movement is hindered by the presence of other particles of the same size, as

$$V_s(r,x) = V_s(r)(1-x)^n \qquad \ldots(5)$$

when $V_s(r)$ is the stokes settling velocity given by Eqn.(4) and x is the volume fraction of particles in the slurry.

Yarandi, Rohatgi and Ray have investigated particle settling in slurry of A356 alloy containing 10,15 and 20 vol% of SiC particles of average sizes of 9μm, 9μm and 14μm respectively[24]. The slurry has been held in a cylindrical crucible isothermally at the holding temperature and the particles are allowed to settle for a given length of time. The furnace is removed from around the crucible and the slurry is quenched in the crucible. The solidified slurry is sectioned and the length of particle denuded zone at the top, has been noted as a function of settling time as shown in .5.

Hanumanth et al have devised a novel four probe electrical resistance measurement technique which records a sharp change in resistance when the boundary between particle free region and the particle containing region moves through the location of the probe[23]. The probe is placed at a known depth and the time taken for the boundary to pass the point of location of the probe is recorded in a x-t recorder. Thus, the variation of particle free region at the top of a slurry for a given time of settling is determined. This method provides a direct measurement of particle free region not interfered by solidification. However, the presence of probe may affect settling locally and may introduce an error. The extent of this problem may be investigated visually by modelling with water containing slurries.

The studies on settling indicates that the finer is the dispersoids and higher is their volume fraction, the rate of settling will be slower. Hanumanth et al used an average dispersoid size of 90μm and a slurry of 0.20 volume fraction of SiC particles settled completely in about 300 seconds resulting in a loosely packed bed of particles at the bottom[23]. At lower volume fraction of particles the settling time is still less. So, it is apparent that a slurry with such large size of dispersoids will have to be stirred all the time till casting. But the results of Yarandi et al[24] carried with relatively smaller average size of SiC particles (9μm or 14μm) show that the initial settling rates are relatively higher but the rates reduce as the volume fraction of particles in the region containing particle, increases as expected according to Eqn. (5). Even after 1 hour of settling, the length of particle free region at the top in a slurry of 0.2 volume fraction of 14μm particles is 14mm while that in a slurry of 0.15 volume fraction of 9μm particles is about 12 mm. Therefore, in slurries containing fine particles (9 to 14μm) it may be possible to do casting by bottom pouring from ladles without any provision for stirring there. After using a large fraction of slurry, the remaining fraction depleted of dispersoids due to settling, could be recycled back to melting/remelting furnace.

2.4 CASTING AND FLUIDITY

The apparent viscosity of slurry increases with solid content and it is therefore expected to reduce the ability of a slurry to flow in mould channels, in comparison to that of the base alloy. However, the ability of an alloy to flow in mould when cooling and solidification are taking place simultaneously, does not depend on viscosity alone but also on the heat transfer conditions deciding the temperature profile and the extent of semi-solid region. While the dispersoids enhance the viscosity of a slurry but those with low thermal diffusivity compared to the base alloy, reduces overall thermal diffusivity of the slurry and therefore, the cooling rate. Even in respect of viscosity, Moon[16] has observed that a semi-solid alloy with externally added particles, has a lower apparent viscosity than that of semi-solid alloy with corresponding total solid content made up of primary phase. The presence of nonmetallic particles of low thermal diffusivity is expected to slow down dendritic growth and prevent coalescence of primary phase particles in stirred slurry. Therefore, externally added particles may counter the tendency to enhance viscosity by formation of a network of primary phases.

Yarandi, Rohatgi and Ray[26] have measured fluidity lengths in sand mould and permanent mould for slurries containing 10,15 and 20 vol% SiC particles in A356 aluminium alloy as shown in Fig. 6 and 7 respectively. It is observed that fluidity length vary almost linearly with temperature upto a limited extent of superheat but beyond it, the results are possibly interfered by chemical reaction of SiC with the base alloy. It is also interesting to note that the fluidity lengths of the base alloy are similar in magnitude to that of slurries and this preliminary investigation shows that there is no significant loss of fluidity length because of presence of SiC particles in the base alloy.

3. PARTICLE SEGREGATION & MICROSTRUCTURE

The micro-structure of cast composite has certain distinct features ----(a) the particles are almost always in the last freezing solid irrespective of their size, (b) particles are also observed in clusters attached to pores, and, (c) primary phases do not nucleate on the particles but the particles may be surrounded by other high temperature phases. All these observations indicate particle segregation and clustering promoting inhomogeneity in particle distribution.

There has been attempts to attribute particle segregation to interaction of particles with solidifying interface. But, it is also possible to inherit particle clusters from earlier steps in processing, i.e., during mixing a slurry, during flow in mould channels and also, during solidification.

When the dispersoids are mixed with molten base alloy, stirring creates a shear field in the slurry. If the gradient of the shear field, i.e., the shear force, in any region of a slurry, is not large enough over the scale of particle size so as to overcome the adhesion force between the particles, there will be stable particle clusters in those regions of the slurry. The particles come in contact during flow and may get attached because of any of the following reasons - physical adhesion, sintering or chemical reaction at the interface. Another source of particle cluster is attachment of poorly wetting particles to bubbles sucked by vortex during stirring or nucleated during solidification because of difference in solubilities of gases between the solid and liquid alloy. The attachment of particles to bubbles make it more difficult for the bubbles to float up and so, are frequently observed in the microstructure.

Slurry flow in mould channels during casting should be investigated with particular attention to particle distribution during flow. There are some investigations of two phase flow in channels in the context of slurry transport. It has been observed that with reducing flow velocity, homogeneous particle distribution in a slurry becomes heterogeneous below a critical value of flow velocity. A further reduction of flow velocity below a second critical level, results in settling of particles at the bottom of channels and their transport, to a limited extent, takes place by tumbling along the channel. These different regimes of flow may be rationalized in terms of Froude Number, F, given by,

$$F = V_f/[2gD(p_s - p_l)/p_l]^{1/2} \qquad \ldots(6)$$

when, v_f is the velocity of flow in a channel of diameter D, of a slurry containing particles of density p_s dispersed in liquid of density p_l. g is the acceleration due to gravity.

Particle segregation may therefore result from two phase slurry flow in mould channels which is much more complex than that investigated for slurry transport. Simultaneous solidification in mould channels and the presence of mushy region are bound to complicate the particle behaviour during flow and there is a need to investigate this problem in the context of casting of slurry so that a Science based mould design could be developed for casting composite components.

It is generally observed that dispersoids like alumina, silicon carbide, graphite etc. are segregated in the interdendritic region of cast microstructure. It was immediately suspected[3] that particle pushing by solidifying dendrites is responsible for locating the particles in the interdendritic region which is last to solidify. The theory of particle pushing as developed by Uhlmann, Chalmers and Jackson, if it reflects the physics of the problem properly, shows that for a particle of a given size, there is a critical velocity of solidifying interface below which the particle is steadily pushed and above, which the particle is eventually engulfed. Subsequent studies[28,29] by others have modified the physics of the problem to some extent but the existence of a critical velocity for a given particle size as described, has been an essential feature of all these theories. The absence of any size specificity in the segregation of particles in cast microstructure led the author to believe that particle pushing by solidifying interface may not be the prime responsible factor for segregation of particles to interdendritic region. Also, it was noted that the surface of dispersoids does not act as sites for heterogeneous nucleation of primary phase when these particles are in constitutionally supercooled region.

Kang, Ray and Rohatgi[30] have investigated one dimensional heat transfer during solidification of Al-Al$_2$O$_3$ slurry and observed that the dispersoid particle with low thermal diffusivity acts as a thermal barrier. Therefore the rate of growth of solid slows down considerably in the region around the particle. These particles are also expected to act as barriers for mass diffusion causing a solute build-up in a region between the solidifying interface and the particle when the solidification front is near the particle and is growing towards it. As a result of solute build-up the liquidus temperature in the region near the particle will come down creating an effect opposite to that of constitutional supercooling and will consequently slow down the growth of solid towards particle as shown in Fig 8. There is a need to carry out detailed investigation to either confirm or deny the arguments presented in this paragraph.

4. CONCLUSIONS

The problems outlined in this article are still being investigated and the conclusions presented in various sections and subsections are tentative. The scientific issues involved in hard core engineering practice of casting composites, are very exciting and needs much more efforts to tackle. The article has been written with the hope that our younger generation of budding technologists in the audience will join in their efforts to unravel and tackle far more exicting science behind the cheapest technique of making cast composite components.

5. REFERENCES

1. S.Ray : in Indian Foundry Journal Special Issue, IFS, Calcutta, India, 1994, pp. 37-44.

2. S.Ray : J. Mat. Sci, 1993, Vol. 28, pp. 5397-5413.

3. V.K. Rai, S.K. Nath and S.Ray : in Proc. IX ISME Conf. on Mech. Engg. S.C. Jain, Ed., Univ. of Roorkee, Roorkee, India, 1994, pp. 407-412.

4. G. Geiger : Ceramic Bulletin. Vol. 70, pp. 212-218.

5. S.Ray : M.Tech Dissertation, IIT Kanpur, India, 1969, pp.49-53.

6. P.K. Rohatgi, R.Asthana, R.N. Yadav and S.Ray : Met. Trans., 1990, Vol. 21A, pp. 2073-2082.

7. L.E. Murr : in Interfacial Phenomena in Metals and Alloys Addison-Wesley Pub. Co., USA, 1975, pp. 280-290.

8. N. Eustathopoulos and B. Drevet : in MRS Symp. Proc. A.H. Carim, D.S. Schwartz and R.S. Silberglitt eds., 1993, pp. 15-25.

9. F.A. Badia and P.K. Rohatgi : Trans. AFS., 1969, Vol. 76, pp. 402-406.

10. A. Alonso, A. Pamies, J. Narciso, C. Garcia-Cordovilla and E. Lous : Scripta Met. et. Mat., 1993, vol. 29, pp. 1559-61.

11. P.K. Rohatgi, S.Ray, R. Asthana and C.S. Narendranath : Mat. Sc. & Engg., 1993, vol. A162, pp. 163-174.

12. S.Ray : To be published in Bull. Mat. Sci. India, 1995.

13. P.K. Ghosh and S.Ray : Ind. J. Tech., 1988, Vol. 26, pp. 83-94.

14. P.K. Ghosh and S.Ray : Trans. JIM, 1988, Vol. 29, pp. 509-519 and 502-508.

15. M.G. Horsten, G.J. Quaak and W.H. Kool : in Proc. Conf. Semi-Solid alloys and Composites, S.B. Brown and M.C. Flemings, eds., MIT, USA, 1990, pp. 359-363.

16. H.K. Moon : Sc. D. Dissertation, MIT, USA, 1990, pp. 359-360.

17. N.E. Elkaddah and K.E. Chang : Mat. Sci. & Engg., 1991, Vol. A144, pp. 221-228.

18. N. Harnby, M.F. Edwards and A.W. Nienow : in Mixing in Process Industries, Butterworth & Co. Ltd., London, 1985, pp. 230-238.

19. P.K. Ghosh and S.Ray : Z. Metallkunde, 1989, Vol. 80, pp. 53-59.

20. P.K. Ghosh and S.Ray : Trans. AFS, 1988, Vol. 88, pp. 775-782

21. P.K. Ghosh, S. Ray and P.K. Rohatgi : Trans. JIM, 1984, Vol. 25, pp. 440-444.

22. R.E. clarity : in Int. Report 89-142, Progress Casting Group, USA, 1989, pp. 8-10.

23. J.F. Richardson and W.N. Zaki : Trans. IChE, 1954, Vol.32, pp.3-15.

24. F.M. Yarandi, P.K. Rohatgi and S.Ray : in Proc. Conf. Semi-Solid alloys and Composites, S.B. Brown and M.C. Flemings, eds., MIT, USA, 1990, pp. 447-465.

25. G.S. Hanumanth, G.A. Irons and S. Lafreniere : Met. Trans., 1992, Vol. 23B, pp. 753-762.

26. F.M. Yarandi, P.K. Rohatgi and S.Ray : Key Engg. Mat., 1993, Vol. 79-80, pp. 91-104.

27. D.R. Uhlmann, B. Chalmers and K.A. Jackson : J. App. Phys., 1964, Vol. 35, pp. 2986-2995.

28. J. Cisse and G.F. Bolling : J. Cryst. Growth, 1971, Vol. 11, pp. 25-39.

29. A.A. Chernov and A.M. Melnikova : Kristallografiya, 1965, Vol. 10, pp. 672-676.

30. C.G. Kang, S.Ray and P.K. Rohatgi: Mat. Sci & Engg., 1994, Vol. A188, pp. 193-199

Table-I : Contact angle and surface properties in selected Metal-Ceramic Systems

System	Holding Temp., °C	Time, mins	Contact angle, θ, in degree	σ_{LV} in mJ/m^2	σ_{SV} in mJ/m^2
Al-SiC	700	10	125 ± 10	860	2950
		160	60 ± 5		
	800	10	110 ± 10	844	
		160	55 ± 5		
Al-Al$_2$O$_3$	600 to 1000	Steady State	$103-.05(T-T_m)$*	$860-.05(T-T_m)$*	900
Al-Oxidized SiC	900	2	120	829	--
		120	50		
Al-graphite	900	1	155	829	--
		30	140		
Al-SiC	1425	-	40 ± 5	730	2910

* T - holding temperature; T_m - Melting point.

Table-II : COMPONENT CASTING FROM REMELTED COMPOSITE INGOTS

Recommendations of Ingot Producers

Manufacturer A	Manufacturer B
1. Premelting:	
⁰ Ingots must be clean, preheated to dry	⁰ Clean dry tools and moulds
⁰ Implements cleaned, coated dried	⁰ All steel tools be coated with chromia, alumina, zircon or boron nitride
2. Melting:	
⁰ Purge furnace with inert gas (dry argon)	⁰ Inert gas may be used
⁰ Continue inert gas flow throughout melting	⁰ In air melting skim surface oxides
⁰ Temp. control mandatory to check chemical reaction and remain below 1400^0F for SiC in 7% Si aluminium base alloy	⁰ Melt should not be taken above 750^0C to avoid chemical reaction
3. Stirring:	
⁰ Required during remelting to counteract particle settling	⁰ Gentle stirring absolutely necessary without breaking the layer on the melt surface
⁰ Avoid turbulence	⁰ Induction stirring is not enough
⁰ Manual or automatic stirring	⁰ If stirring is interrupted, start again and continue for 10 minutes in the worst case
4. Pouring:	
⁰ Avoid turbulence	⁰ Bottom pouring with continued stirring recommended
⁰ Ceramic foam filters or screens may be used	⁰ Woven fibre filters or preheated ceramic foam filters may be used
5. Casting Design:	
⁰ Feeding Distance should be about 65% of that for the base alloy; additional riser may be required	⁰ Enlarge gates and sprues by 25% over that for the base alloy

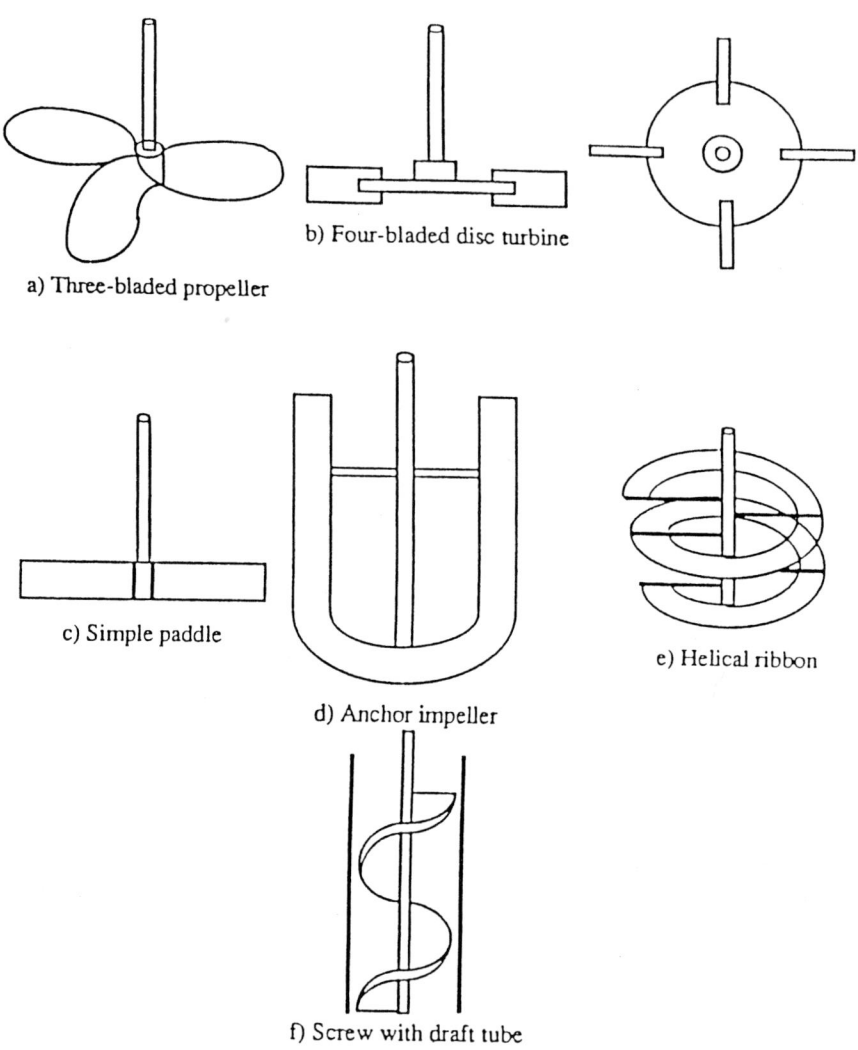

Fig. 1   Schematics of basic stirrer designs.

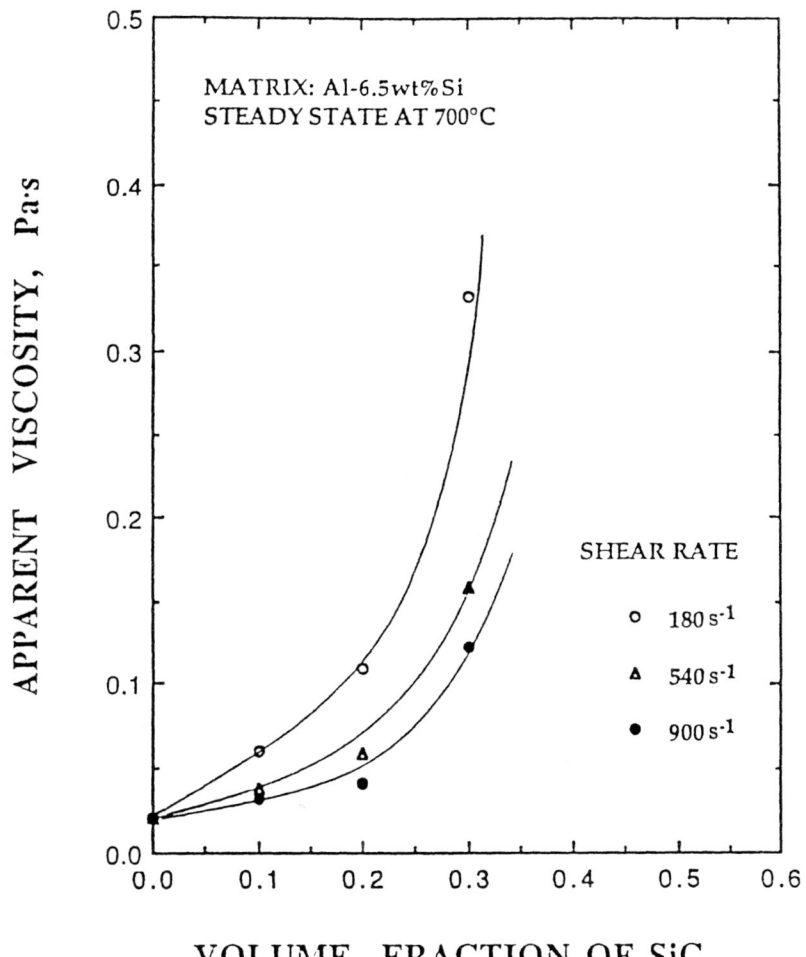

Fig. 2 Variation of steady state apparent viscosity with volume fraction of silicon carbide particles in slurry agitated at different shear rates.

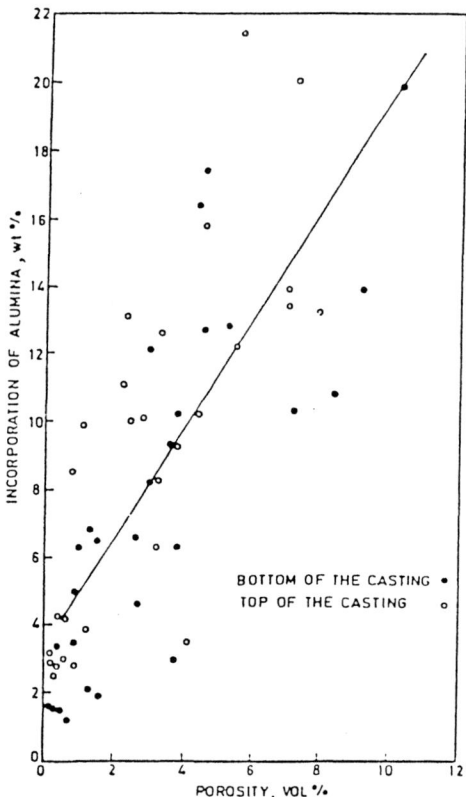

Fig. 3   Variation of porosity with alumina content in cast Al-Al$_2$O$_3$ Composites.

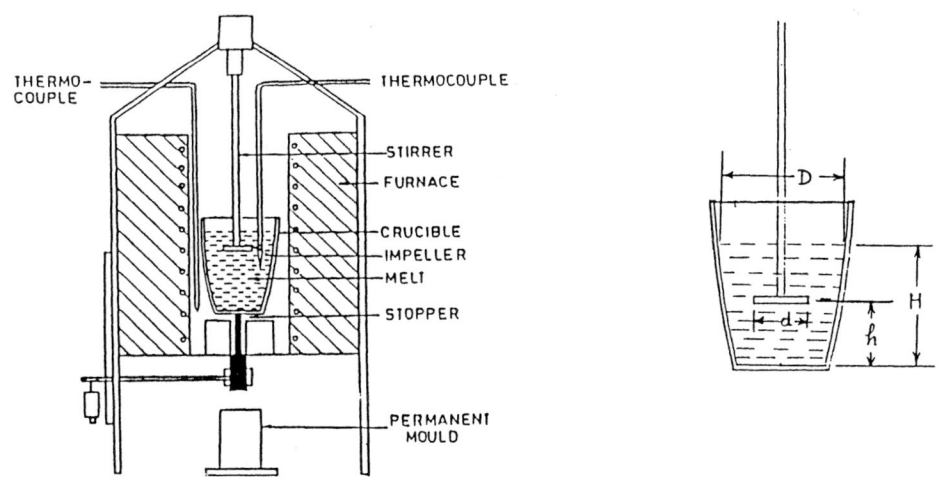

Fig. 4 Schematic Set-up used used for stircasting and compocasting and dimensions of crucible and stirrer.

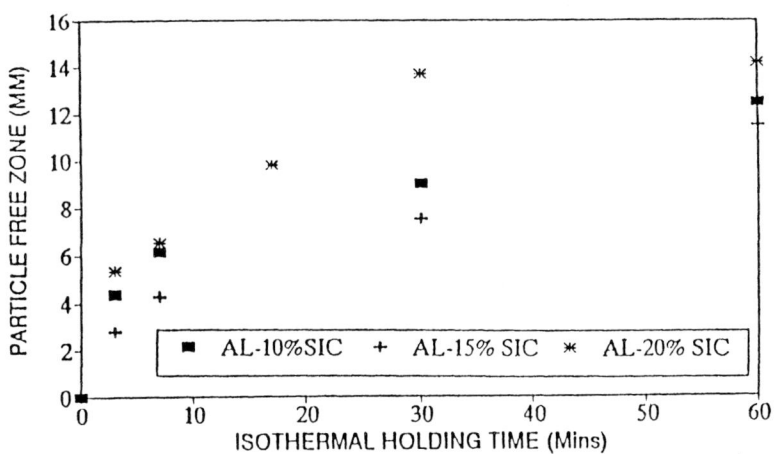

Fig. 5 Variation of particle free region at the top of a slurry of A356 aluminium alloy containing 10, 15 and 20 vol% of silicon carbide particles.

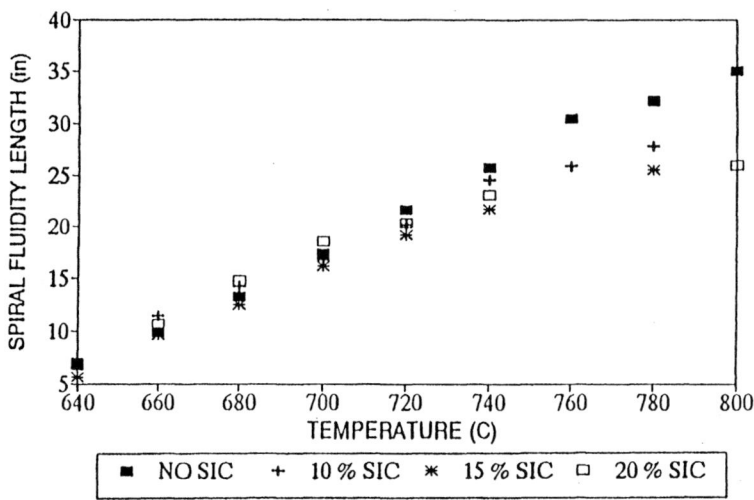

Fig. 6   Sand mould spiral fluidity lengths for slurries containing 0, 10, 15 and 20 vol% of SiC particles in A356 aluminium alloy.

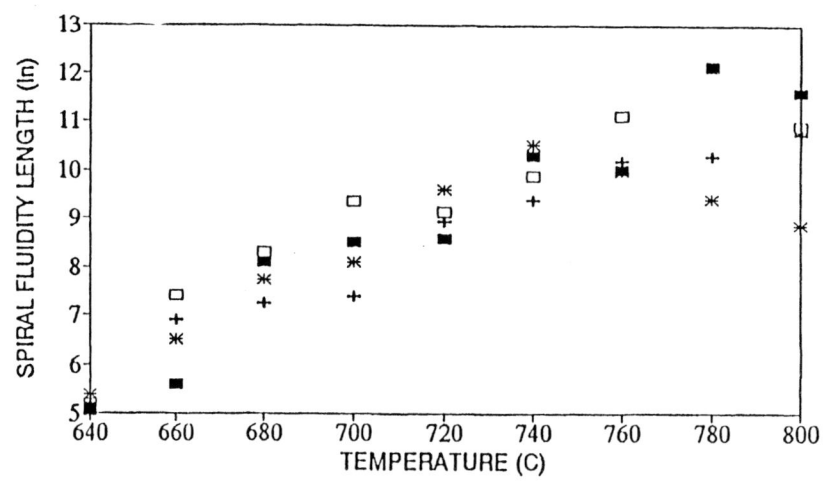

Fig. 7   Permanent mould spiral fluidity lengths for slurries containing 0, 10, 15 and 20 vol% of SiC particles in A356 aluminium alloy.

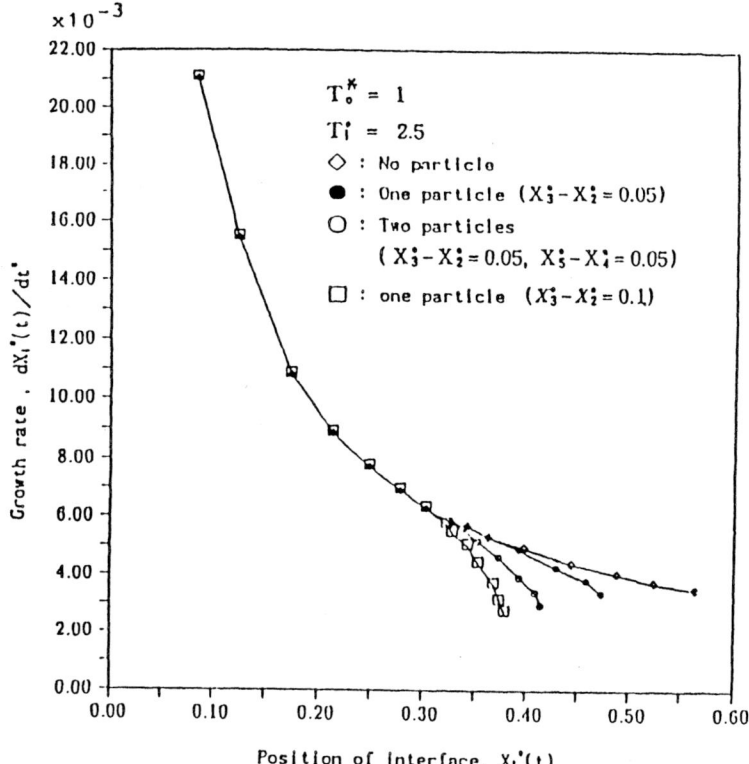

Fig. 8 Variation of growth rate with position of solidifying interface in the absence and presence of alumina particles of different sizes in aluminium melt.

METAL MATRIX COMPOSITES FABRICATED BY PRESSURE ASSISTED INFILTRATION OF LOOSE CERAMIC POWDER

Mohamed A. Taha and Nahed A. El-Mahallawy

Department of Design and Production Engineering
Faculty of Engineering, Ain - Shams University
Abaseia, Cairo - Egypt

ABSTRACT

A composite fabrication method has been developed in which a pressure is applied to infiltrate loose ceramic particles by liquid metal.

The pressure applied has been exerted by two different techniques: centrifugal and squeeze castings. Both techniques are found successful in preparing Al / Al_2O_3 MMC's.

In this method the alumina powder followed by a rod of aluminium are inserted in a tube which is heated to a temperature above the melting point of the alloy. In centrifugal casting the tube is rotated around an axis perpendicular to that of the tube and a centrifugal force is induced which acts on the liquid metal. In squeeze casting a squeezing force is applied on the liquid metal using a similarly heated plunger. The composite is formed due to the infiltration of the liquid metal under the action of the force through the powder interstices.

The infiltration mechanisms in both techniques are found different. While the formation of the composite was in the direction of the squeezing force, it was in a direction opposite to the centrifugal force. However four cases of infiltration are obained in both techniques: no infiltration because the pressure applied was lower than the back pressure due to surface tension, partial infiltration with remaining metal above the alumina powder due to low pressing time, full infiltration due to enough metal charge and enough time and partial infiltration due to non-sufficient metal charge. The infiltration mechanisms in both cases are suggested in view of the application of different processing conditions, Al_2O_3 different particle size and using a commercial aluminium and an Al-Si alloy.

The structure features of the composite namely the particle distribution, the metal / ceramic interface and the soundness are studied. A comparison between the structures obtained in both techniques is made. In both cases, a high volume fraction of Al_2O_3 ranging between 50 and 65% is obtained.

1. INTRODUCTION

Due to the increased use and importance of metal matrix composites (MMCs), their fabrication techniques have been subjected to a continuous development during the last few years. A variety of methods for producing MMCs have recently become available. Among these methods, fabrication using liquid state processing (casting) has a potential advantage of achievement of a near - net shape product in a simple and cost effective manner. The wide selection of materials and the good bonding between particles and matrix are also achieved.

In terms of both processing and properties, the main concern that must be adressed is that of uniform particle distribution [1]. While this is achieved in pressure assisted infiltration of preforms (fibres or particles) by liquid metal, other casting techniques encounter several problems such as floatation and agglomeration, extensive interfacial reactions or segregation of the second phase in the matrix and the ceramic reinforcement [1]. However, preforms have limitations in their use and their volume fraction. Another important problem is the poor wettability between the liquid metal and the ceramic reinforcement [2,3]. In the infiltration techniques, the pressere is applied to overcome such problem especially in the case of Aluminium composites. The pressure applied, should be sufficient to guarantee perfect infiltration of the liquid metal into the inter-particle / fibre recess and to achieve a better interfacial continuity.

Recently, the authors have developed two pressure - assisted techniques to infiltrate loose ceramic powder instead of a preform. One of these techniques is the application of centrifugal force to infiltrate the liqiud metal between the loose powder [4,5]. The other technique is applying a squeezing force for the infiltration [6,7]. By these techniques, successful Al-Al_2O_3 particulate MMCs have been prepared with high volume fraction of Al_2O_3 particles (50-70%). These composites have proved to be promising for their good wear properties [8].

In this paper, a comparison is made between the two techniques and their infiltration mechanisms are discussed. The main structure features obtained in the composites prepared are briefly presented.

2. EPERIMENTAL TECHNIQUES

The composites were prepared from commercial Al and Al-12Si-2Mg as matrix and alpha Al_2O_3 (alumina) powder with different particle size as reinforcement. The measured bulk density of the Al_2O_3 loose powder was 2.9 g/cm3.

In the centrifugal casting a cylindrical rod of Al-12Si-2Mg was inserted in steel tube on top of a quantity of loose alumina. The system was contained in a cylindrical insulated and sealed capsule (Fig.1-a) which was heated to melt and superheat the alloy to 100 K above the liquidus as well as preheating the powder. The tube was then located in a centrifugal set up and was rotated around a horizontal axis (700 - 2000 rpm), Fig.1-b. The rotation period was enough to infiltrate the liquid metal through the powder under the action of the centrifugal force as well as to solidify it. Composite rods 12 mm in diameter were formed.

In squeeze casting, a cylindrical metallic rod (commercial Al) was also inserted on the top of a quantity of of loose alumina inside an internally coated steel die, Fig.2-a.
The system including a punch above the metal was also heated, then a pressure was applied using a hydraulic press immediately when the melt temperature was 100 K above the liquidus, Fig.2-b. The pressure was applied until complete infiltration of the liquid metal and its solidification was completed. The pressures in the range frm 65 to 180 MPa were applied. Composite rods 15 mm in diameter were formed.

3. **RESULTS AND DISCUSSION**

3.1. Infiltration Cases

The extension of infiltration was examined on longitudival sections made on the composite samples. In both centrifugal and squeeze casting techniques, the following four cases of infiltration as sketched in Fig.3 (centrifugal casting) and Fig.4 (squeeze casting) were obtained depending on the process conditions:

1) No infiltration; where no liquid metal was able to go through the intersticies between the powder and therefore no composite was formed.
2) Partial infiltration with remaining metal above the powder. In centrifugal casting the composite is formed at the far end of the powder length and it is separated from the remained metal by the un-infiltrated powder part, Fig.3. In squeeze casting, the composite was formed in the upper part of the powder length adjacent to the remaind metal, Fig 4.
3) Full infiltration; where the liquid metal was able to infiltrate the full length of the powder so that a composite was formed in the whole powder length and a remained metal was found above the composite specimen.
This case happened when enough initial metal was charged and the infiltration conditions (time and pressure) were suitable.
4) Partial infiltration; with no remaining metal due to

insufficient intial metal charged. The location of the composite formed is similar as described in case(2).

3.2. The Acting Pressure and Infiltration length

The pressure at the metal / powder interface is the acting infiltration pressure. The effect of this pressure on the infiltration length is studied.

1) Centrifugal Casting

In centrifugal casting the acting pressure decreases as the mass of liquid metal above the powder decreases during infiltration. It is calculated from the equation: mw^2R/A, where m is the molten metal mass above the powder changing during infiltration; w is the angular velocity, R is the mean redius of rototion of the molten metal and A is the specimen cross - section area.

Fig. 5 shows the effect of acting pressure on the composite length for specimens prepared using different initial length of metal and speeds, while the powder length and grain size were kept constant at 40 mm and 47 um respectively. The results of different initial metal lengths (Lm) fall on the same curve for a constant powder grain aize. With increasing pressure, the composite length increases up to a maximum value of 40 mm close to the initial powder length. This length is reached at a pressure of 130 KPa. With higher pressures, the composite length remains constant. Also with adding more initial metal charge, no increase in the composite length was obtained. This indicates that the powder is not displaced significanly during infiltration. Such observation is also supported from measurements of alumina volume fraction in the solidified composite specimens, which was found similar to that of the loose powder before infiltration (57% in this case). As $Al-Al_2O_3$ system is nonreactive [9], a straight line relationship between pressure and composite length is obtained before the maximum composite length is reached as shown in Fig.5.

In the case of partial infiltration where the composite length is less than the maximum value (40 mm), the composite start forming at the far end of the mould and then progresses in the direction opposite to that of the centrifugal force, Fig.3. The partial infiltration can be explained on bases of variation in the acting pressure during infiltration. At the begining of rotation the mass of liquid metal was large enough to give an acting pressure sufficient for infiltration. During the process, this mass decreases and the resulting pressure decreases until the infiltration progress stops. As indicated in Fig.5, a minimum pressure of 1 0 kPa is necessary for full infiltration. This value depends on the grain size of the powder, the rotating velocity and the distance from the centre of rotation. Other factors, such as type and surface conditions of powders as well as alloy type and superheat are also expected to affect the infiltration [3,10].

2) Squeeze Casting

In this case, the acting pressure at the metal / powder interface is constant during infiltration. It is calculated by deviding the squeezing force by the specimen (plunger) cross - section area.

In order to study the effect of the acting pressure on infiltration for this case, the specimens were prepared with similar lenghs of initial powder and metal charge which had a constant value of 40 mm for all experiments conducted. Different Al_2O_3 grain sizes (80,95 and 115 um) and different squeezing pressures (65,90,130 and 180 MPa) were applied.

The results are remarkable in such that the composite length was always slightly higher than the initial powder length. Such difference in lengths increases with increasing either presure applied and/or Al_2O_3 particle size as indicated from Fig.6 and 7 respectively. However the value of such increase did not exceed 1.8 mm even for the highest pressure and largest grain size used. This is correlated with measurements of interpartcle spacings (between Al_2O_3 Particles) which were found to be higher as the pressure was increased.

3) Comparison between Centrifugal and Squeeze Casting

Generally, the present experiments show that the pressure needed for full infiltration is less in centrifugal casting than in squeeze casting. In centrifugal casting a pressure of 120 kPa was enough to achieve specimen full infiltration for Al_2O_3 size of 47 um. The lowest squeezing pressure used to achieve specimen full infiltration even for a much coarser Al_2O_3 grain size (115um) was in the order of 65 MPa. Similar conclusion is also obtained in a comparison between the present work on centrifugal method and a previous work [3,10] on squeeze infiltration on 2014 Al / SiC particles with similar particle size, Fig.8.

3.3. Propagation of Infiltration with Time

In sets of experiments, the infitration time was changed so that partial infiltration was obtained as in case (2) and the progress of infiltration was followed.

1) Centrifugal Casting

In this case, the experiments were achieved on partially rotated specimens (rotation and then stop). The results show that the composite starts to build at the far end of the specimen and progresses towards the centre of rotation or towards the metal / powder interface (opposite to the pressure direction). As shown in Fig.9, it is seen that full infiltration is achieved in less than two seconds. However, more than half a minute was necessary for full densification of the specimens. In this case the composite is formed with a maximum length almost equal to that of the unconsolidated powder before infiltration.

2) Squeeze Casting

In these experiments, the squeezing time (duration of pressure application) was varied while other processing conditions were kept constant. The results show that the composite forms first at the metal / powder interface and propagates into the powder in the direction of pressure application. As shown in Fig. 9, full infiltration was achieved at 25 seconds. This is a much longer time in comparison with few seconds obtained in centrifugal casting. This is probably due to the dynamic action of centrifugal force which could shorten the time. In a previous work by Asthana et al [3] on infiltration of 2014 Al alloy into porous platelet SiC compacts using countergravity N_2 much lower squeezing pressure of 800 KPa, a much longer infiltration time was needed (3 to 30 minutes) depending on the surface conditions of the powder.

3.4. Infiltration Mechanisms

1) Centrifugal Casting

Fig. 10 proposes a schematic drawing of the infiltration mechanism where two opposite forces act on each element of the molten metal : a) the centrifugal force (Fm), which is forcing the molten element in the infiltration direction, and is depending on its mass, radius of rotation and rotational speed, and b) the resistance (R^*) of the ceramic powder due to surface tension, and particle size. The figure shows that a cumulative force [Fm] acts on the surface of the powder which helps starting the infiltration. Once the molten metal enters the interstices between the powder particles, it may behave as short streams which travel between the particles and continue their way due to the centrifugal force Fm. If this force is less or equal to the resistance (R^*), the streams will stop. There is usually a critical radius below which the infiltration stops. The minimum pressure needed for infiltration (representing Fm) is evaluated for the different cases. It falls between 50 and 90 kPa. These values are less than the threshold pressures reported for squeeze casting of 2014 Al alloy through SiC particles (100 to 500 kPa) [3,10]. They are also lower than those reported for vacmuum infiltration through bundles of SiC coated carbon fibers [11].

2) Squeeze Casting

Fig.11 proposes a schematic drawing of the infiltration mechanism in squeeze castinge. As the pressure is applied on the liquid metal, a resistance to infiltration or back ressure is exerted due to molten metal surface tension [11]. As the pressure applied on the liquid exceeds this back pressure the liquid penetrates the metal / powder interface and starts to flow between the particles i.e propagation of infiltration front with time in direction of pressure. It is noticed that the front velocity decreases with time. When the front reaches the bottom of the mould such velocity approaches zero. During the propagation of the front, it was noticed that a densification process occurs simultaneously on the same side and a moving densification front is said to occur. This result is obtained from density measurements on partially infiltrated samples with different lengths. It can be deduced that the densification front velocity is almost constant while the infiltration front velocity decreases. The constant velocity of densification front is deduced from the fact that both fronts meet at the bottom of the die leading to uniform density along the whole section of the composite.

3.5. Particle Distribution

The nature of the infiltration mechanisms described was reflected on the uniformity of the particle distribution.

In centrifugal casting, as no significant powder displacement occured during infiltration, an almost uniform distribution of Al_2O_3 particulates was achieved, Fig.12. This is observed in both longitudinal and transverse sections. The interparticle spacing was found with almost constant value in the specimen. Amongst the processing variables applied, only changing powder grain size was found to cause variation in the interparticle spacing [5].

Microscopic observations on squeeze cast specimens, indicate that although the distribution of Al_2O_3 in the transverse section was almost uniform, it was not uniform in the longitudinal section, Fig.13. While at the specimen top the particles are relatively far from each other, they are closer to each other towards the bottom. A gradual increase in the interparticle spacing was observed from bottom to top of the specimen [6]. Generally the interparticle spacing obtained in squeeze casting was relatively larger than in centrifugal casting.

While an almost perfect particle distribution was obtained in centrifugaly cast specimens, few instances of Al_2O_3 depleted areas have been noticed in some squeeze cast specimens. Two types of such defect, namely transverse chaneling and localized disruption of compacts have been observed [6].

4. CONCLUSIONS

1) Centrifugal and squeeze casting of loose particles are successful methods for producing composites with high ceramic volume fraction. Due to its dynamic nature, the acting pressure in centrifugal casting was much lower than the squeeze pressure.

2) Proposed infiltration mechanism for centrifugal casting expects that the composite length is unlimited, given that the metal fluidity is not altered. The composite formation starts from the mould far end and progresses towards the centre of rotation opposite to the direction of the centrifugal pressure.

3) Proposed infiltration mechanism for squeeze casting expects that the composite formation starts from the metal / powder interface and the infiltration front progresses towards the mould far end in the direction of pressure. A densification front follow simultaneously the infiltration front until meeting at the far mould end.

4) Almost uniform particle distribution is achived by centrifugal casting while in squeeze casting it is not uniform the presence of some particle depleted areas.

REFERENCES

1. D.M.Stefanescu, in M.A.Taha and Nahed A. El-Mahallawy (eds) Advances in Metal Matrix Composites, Key Engineering Materials, Vol. 79-80, Trans Tech. Publ. Ltd, Switzerland, 1993, p. 75-90.

2. A.Banerji, P.K.Rohatgi and W.Reif, Metall, 38 Jahrgang, Heft 7 (1984) 656-661.

3. R.Asthana and P.K.Rohatgi, Z.Metallkunde, 83, H 12 (1992) 887-892.

4. N.A.El-Mahallawy, M.A.Taha, A.K.El-Kharbotly, A.F.Yousef and W.Reif, Cast Metals, 7, number 3 (1994) 175-183.

5. N.A.El-Mahallawy, M.A.Taha, A.K.El-Kharbotly, A.F.Yousef and W.Reif, Cast Metals, 7, number 3 (1994) 185-190

6. M.A.Taha, N.A.El-Mahallawy and M.Abdel-Hamid, to be published.

7. M.A.Taha. M.H.Abdel-Latif, N.A.El-Mahallawy and H.Abdel-Hakim, Proc. 4th Int. Conf. on Production Engineering and Design For Development, Cairo, December 1993, p.109-119.

8. M.A. Taha, N.A.El-Mahallawy, M.Abdel-Hamid and H.A.Hanna in D.M.Stefanescu and S.Sen (eds), Proc. 2nd Int. Conf. on Cast Metal Matrix Composites, October 4-6, 1993, Tuscaloosa, Alabama, American Foundryman Soc., Des Plaine, Illinois, USA (1994) 335-345.

9. S.Y.Oh, J.A.Cornie and K.C.Russel, Met Trans. A, 20A (1989) 527-532.

10. R.Asthana and P.K.Rohatgi, in M.A.Taha and Nahed A. El-Mahallawy (eds), Advances in Metal Matrix Composites, Key Engineering Materials, Vol. 79-80, Trans. Tech. Publ. Ltd, Switzerland, 1993, p.47-62.

11. X.Zhenhai, M.Zhiying and Z.Yaohe, Z.Metallkunde, 82, H 10 (1991) 766-768.

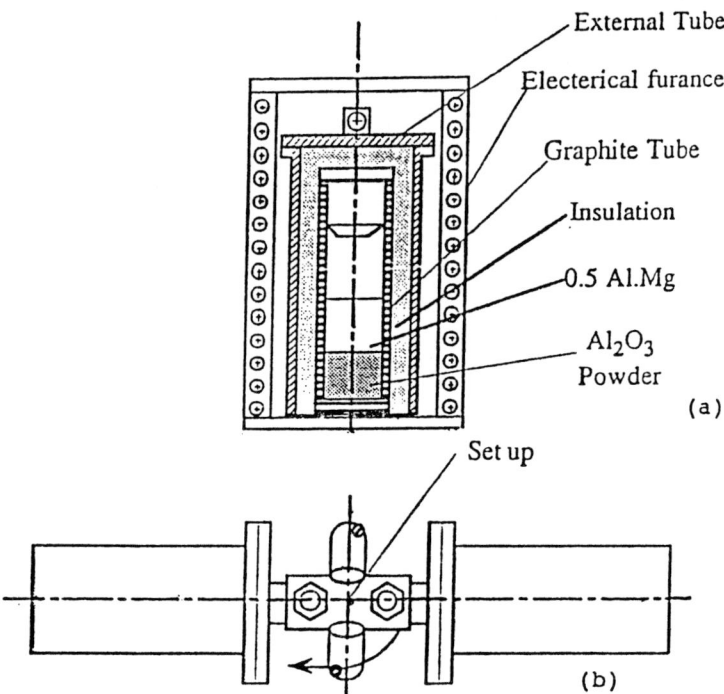

Fig.1. Sketch showing (a) principle of centrifuged casting technique and (b) mould assembly.

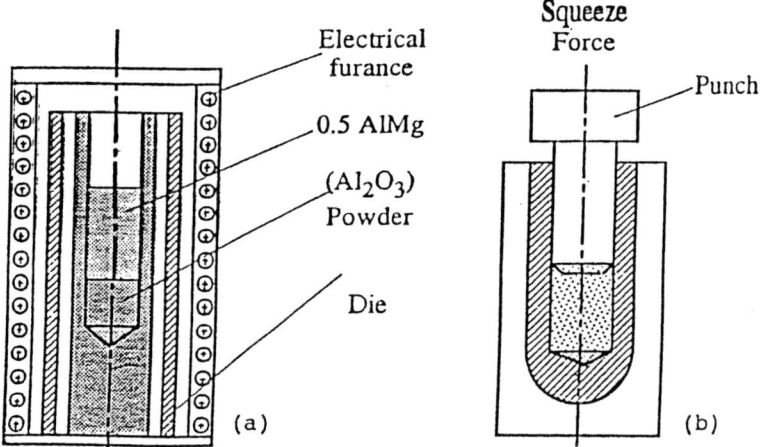

Fig.2. Sketch showing (a) principle of squeeze casting technique and (b) die assembly.

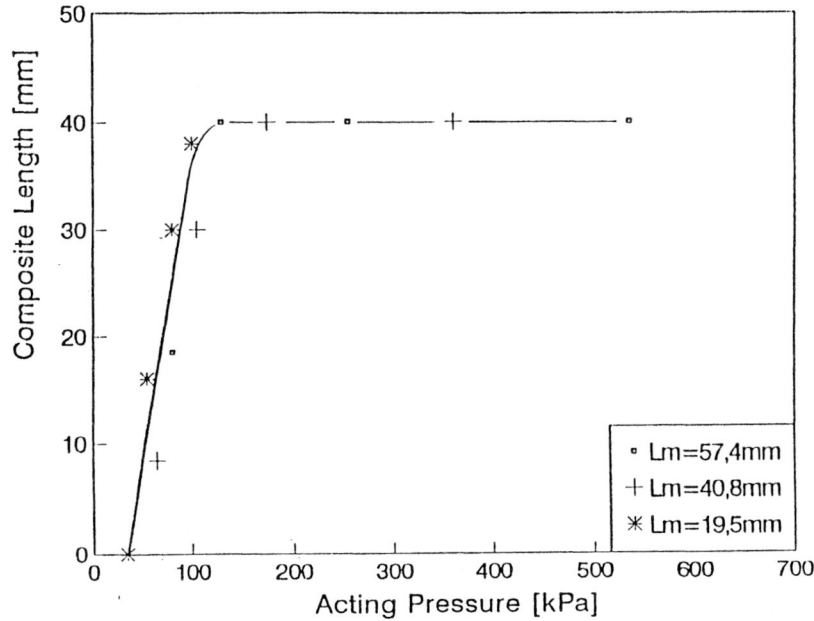

Fig.5. Effect of acting pressure on infiltration length.

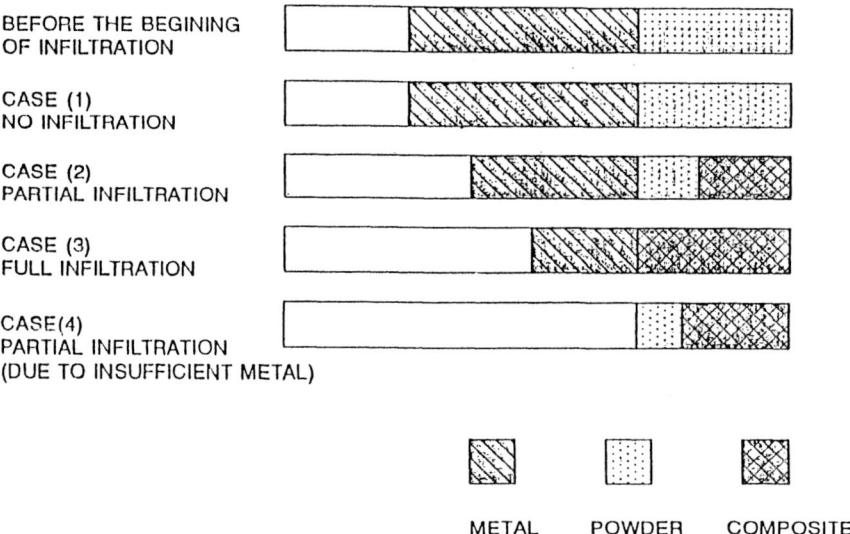

Fig.3. Sketch showing the different infiltration cases.

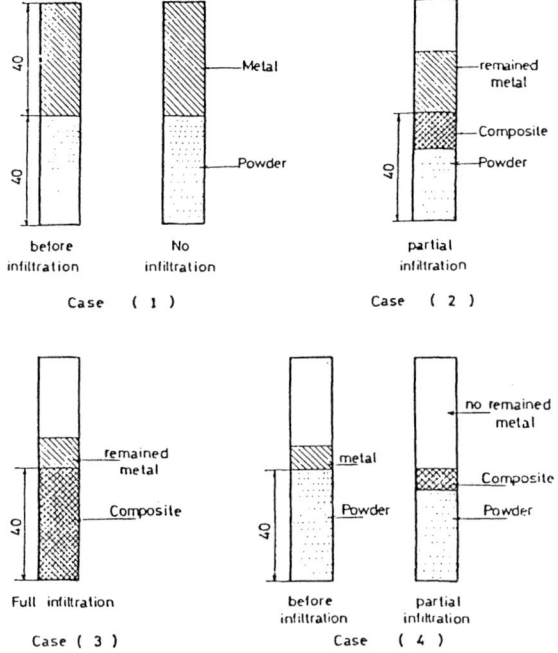

Fig.4. Sketch representing the cases of infiltration.

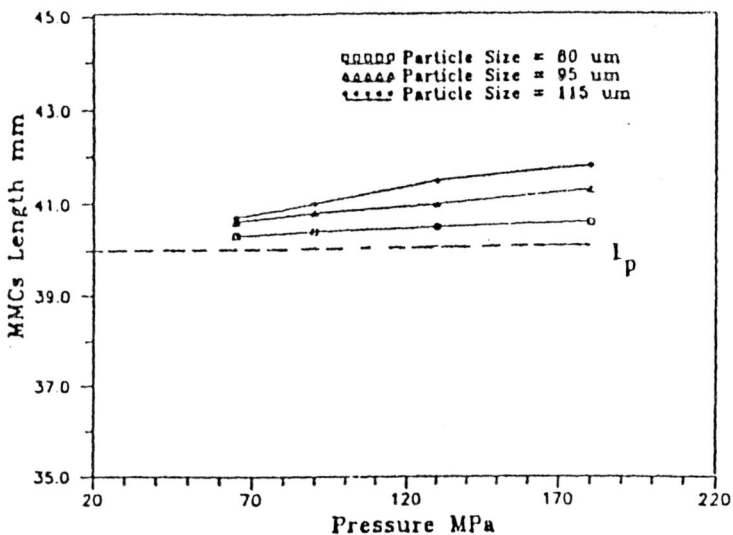

Fig.6. Effect of squeeze pressure on MMC length

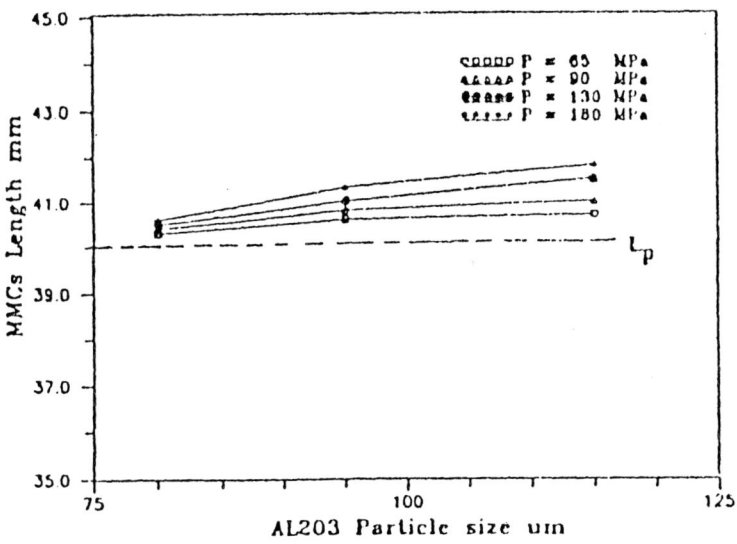

Fig.7. Effect of Al_2O_3 particle size on MMC length

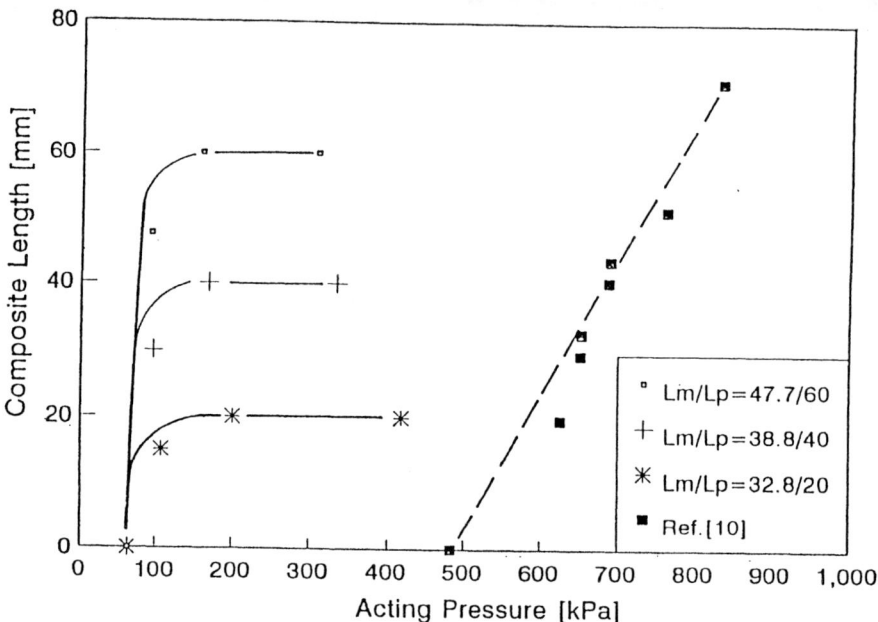

Fig.8. Effect of acting pressure on composite length for different powder length.

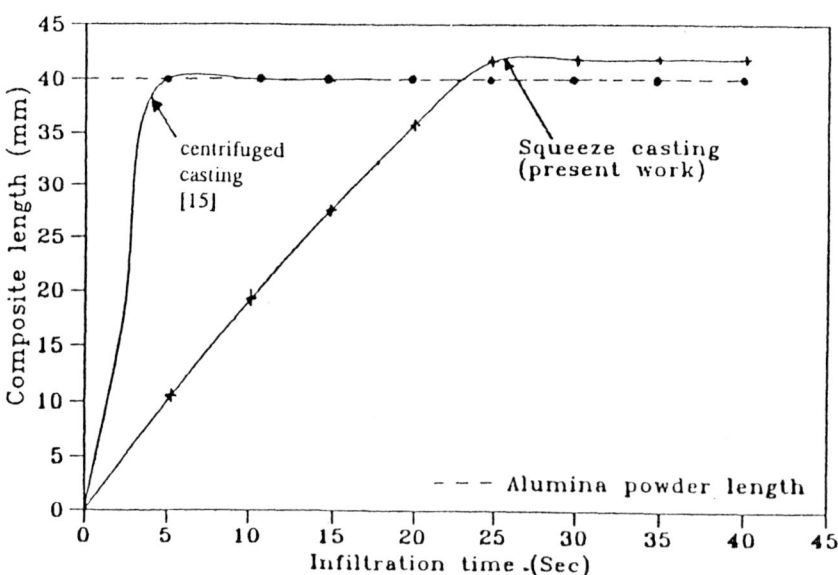

Fig.9. Composite length versus the infiltration time.

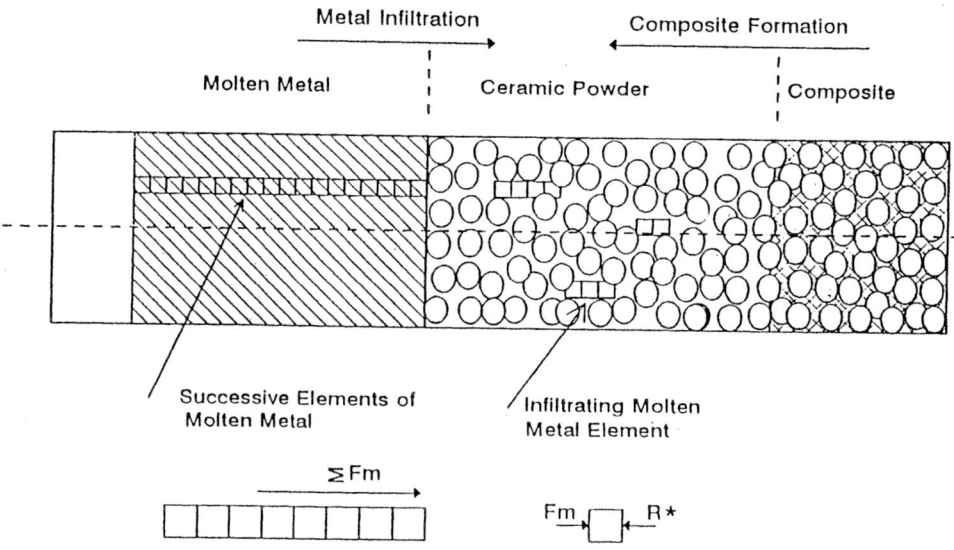

Fig.10. Sketch repressenting the infiltration mechanism where (Fm) is the centrifugal force acting on the molten metal, (ΣFm) is the accumulated force, (R*) is the resistance force to the infiltration.

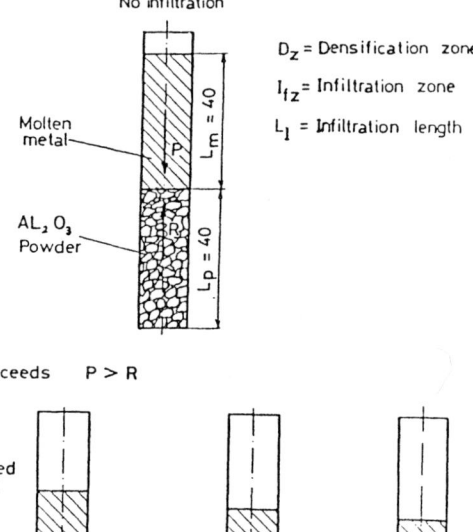

Fig.11. Sketch repesenting the infiltration machanism, where P infiltration pressure acting on the molten metal, R is the resistance force.

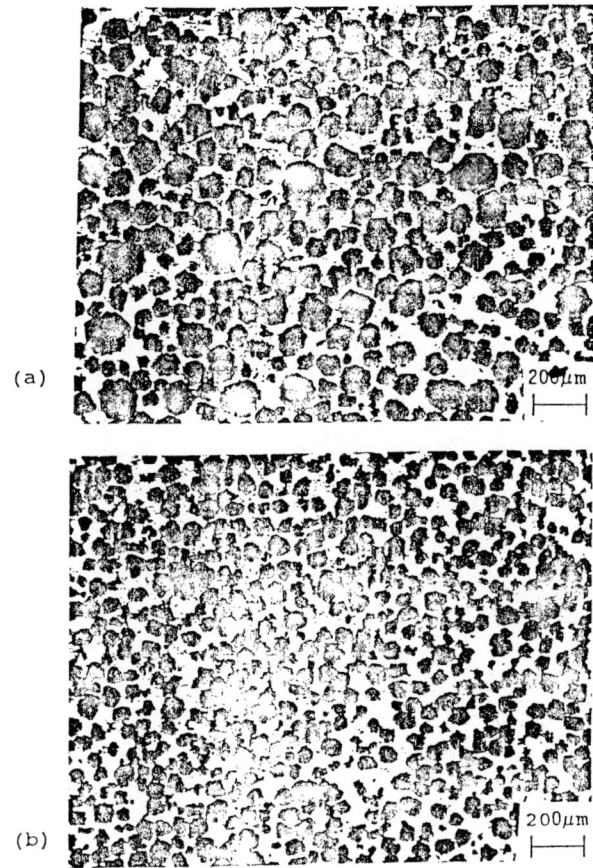

Fig.12. Typical photomicrographs for Centrifugally cast specimens with different Al_2O_3 size:
(a) coarse: 89.8 Um, (b) medium: 47.9 Um.

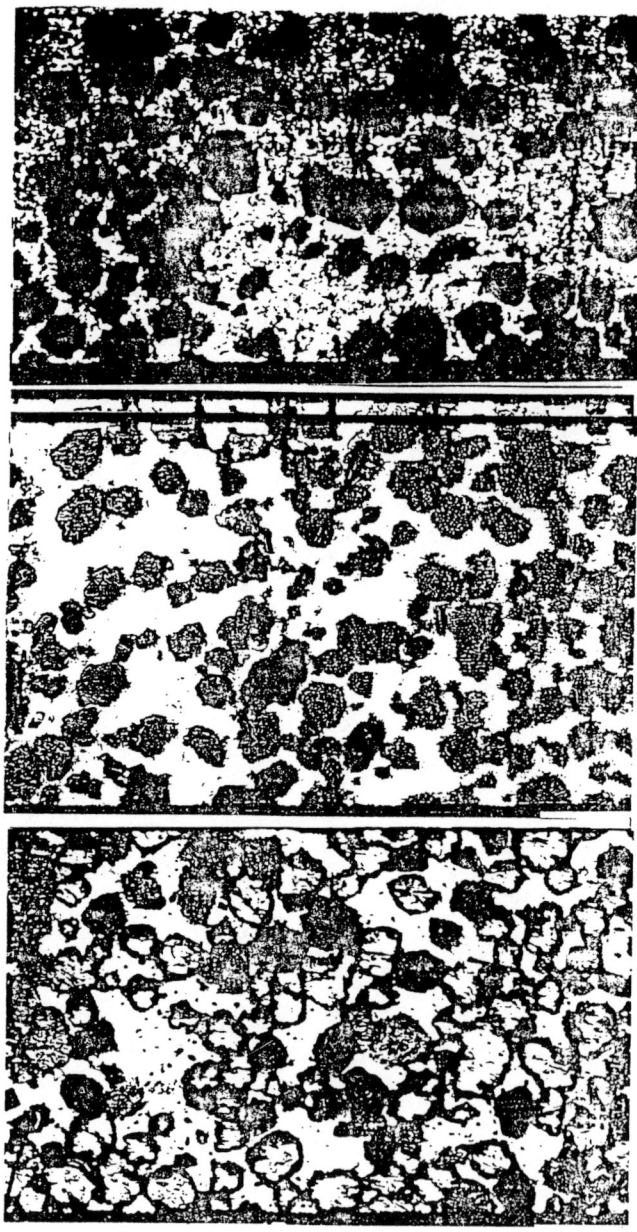

Fig.13. The particle distribution along the length of squeeze cast sample with particle size 95 um:

Effect of Type of Processing on the Microstructural Features and Mechanical Properties of a SiC Reinforced Al-Cu Metal Matrix Composite

M. Gupta, L. Lu, M.O. Lai and A.S. Ee
Department of Mechanical and Production Engineering
National University of Singapore, 10 Kent Ridge Crescent, Singapore 0511

Introduction

The unified combination of metals and ceramics, commonly known as metal matrix composites (MMCs), have fueled the research activities globally due to their potential to serve a spectrum of applications [1-2].

The suitability of MMCs as a viable replacement of the conventionally used monolithic materials, however, depends on the acquisition of scientific understanding in order to synthesize them with consistent reproducibility in microstructure and mechanical behavior and their ability to exhibit enhanced performance based cost effectiveness in real time application[1, 3].

The enhanced performance from these rather unique materials depends on a careful selection of processing technique/parameters, matrix material and the reinforcing phase. Liquid phase processes, for example, beside their disadvantages are still the most economical processes to synthesize MMCs [4]. However, the commercial utilization of liquid phase processes is still very much limited as a result of limited information available in open literature regarding the synthesis of these materials. Moreover, MMCs processed using liquid phase processes exhibit coarser microstructure and non uniform distribution of SiC particulates as a result of sluggish solidification velocity resulting into relatively inferior mechanical properties. In order to circumvent the problems associated with liquid phase processes, partial-liquid phase processes have been investigated and reported to exhibit an improved combination of microstructural characteristics and mechanical properties [3, 5]. The partial-liquid phase processes, however, are met with limited success and further efforts are required in order to gain further understanding so as to synthesize MMCs with an improved microstructure and enhanced mechanical properties.

Accordingly, in the present study metal matrix composites were synthesized using liquid phase and partial-liquid phase techniques in order to gain further insight into the microstructural evolution as a result of change in certain processing steps. The as-processed composites thus obtained were microstructurally characterized using scanning electron microscopy and tensile tested using an automated servohydraulic Instron testing machine. The mechanical properties thus obtained were correlated with the processing associated microstructural features of the composite materials.

Experimental Procedure

Materials
The nominal composition of the matrix alloy used in the present study was (in wt. %): 2.0Cu - Al (bal.). Silicon carbide (α-SiC) particulates with an average size of 23 µm were selected as the reinforcement phase.

Processing
The synthesis of the metal matrix composites used in the present study was carried out using conventional casting and modified rheocasting techniques. The synthesis of MMCs using conventional casting was carried out according to the following procedure. The metal ingots, prior to melting, were properly cleaned to eliminate surface impurities. The cleaned metal ingots were melted to the desired superheating temperature. SiC particulates, preheated to

Inorganic Matrix Composites
Edited by M. K. Surappa
The Minerals, Metals & Materials Society, 1996

900 °C, were then added into the molten metal stirred using an impeller. The composite melt thus obtained was poured into cylindrical steel molds (25 mm diameter and 178 mm height). In all the cases, stirring time of SiC particulates in the melt was maintained between 10 and 15 minutes. Regarding the rheocasting of MMCs, the synthesis process involved: superheating the properly cleaned metal ingots in a graphite crucible, addition of preheated SiC particulates in the liquid metallic melt and stirring of the composite mixture in the liquid phase regime and the two phase regime in order to achieve the uniform distribution of SiC particulates in the metallic matrix. The composite material thus obtained was allowed to solidify in the crucible and was subsequently remelted in the crucible followed by casting into cylindrical steel molds.

For the purpose of comparison, the base alloy was cast under similar processing conditions as described for conventional casting process.

Quantitative Assessment of SiC Particulates

Quantitative assessment of SiC particulates in the composite samples was carried out using a chemical dissolution method. This method involved: i) measuring the mass of composite samples, ii) dissolving the samples in hydrochloric acid, followed by iii) filtering to separate the ceramic particulates. The particulates were then dried and the weight fraction determined [5].

Density measurement

Density measurements were carried out in order to ascertain the volume fraction of porosity in the Al-Cu matrix. Density measurements were carried out using Archimedes' principle following the procedure as discussed in Reference [4].

Aging Studies

Aging studies were carried out in order to obtain the peak hardness temperature and time conditions for the unreinforced, conventionally cast and rheocast metal matrix composites. Specimens (25 mm diameter x 7 mm height) taken from as-processed rods were solutionized for one hour at 490 °C, quenched in cold water and aged isothermally at 160 °C for various intervals of time. Rockwell B hardness measurements were made using 1.58 mm diameter steel ball indentor with a 100 kg load.

Microstructural Characterization

Microstructural characterization studies were conducted on the as-processed samples in order to investigate the distribution of SiC particulates and the presence of porosity. Particular emphasis was placed to examine the precipitation behavior and segregation of alloying elements in the interfacial region between the Al alloy matrix and ceramic particulates.

Microstructural characterization studies were primarily accomplished using a JEOL scanning electron microscope equipped with EDS [Energy Dispersive Spectroscopy]. The composite samples were metallographically polished prior to examination. Microstructural characterization of the samples were conducted in both etched and unetched conditions. Etching was accomplished using Keller's reagent [0.5 Hf - 1.5 HCl - 2.5 HNO_3 - 95.5 H_2O].

X-Ray Diffraction Studies

X-ray diffraction study was carried out on the unreinforced, conventionally cast and rheocast specimens using an automated Shimadzu XD-D1 diffractometer. Thin samples were exposed to CuK_α radiation (λ = 1.5418 A°) using a scanning speed of 8 deg min^{-1}. A plot of intensity versus 2Θ was obtained, illustrating peaks and the corresponding d spacings at different Bragg angles. The values of d thus obtained were matched with standard values for aluminum and other phases provided in reference [6].

Mechanical Behavior

The smooth bar tensile properties were determined on the peak aged monolithic and composite specimens following ASTM standard E8-81. Tensile tests were conducted using an automated servohydraulic Instron testing machine on 4 mm diameter specimens using a crosshead speed of 0.254 mm per minute.

Results

Macrostructure

Macrostructural characterization conducted on the as-processed, machined and polished conventionally cast specimens revealed the presence of macropores and the macrosegregation of SiC particulates. These features, however, could not be detected on the machined and polished surfaces of unreinforced and rheocast composite specimens in the as-processed condition.

Quantitative Assessment of SiC Particulates

The results of acid dissolution experiments are summarized in Table 1. The weight percentages of SiC particulates was estimated to be approximately 13.9 % for conventionally cast composite specimens and 10.9 % for the rheocast composite specimens.

Table 1. Results of the acid dissolution tests and volume percent porosity determination.

Matrix	Condition	Reinforcement Size	Weight % SiC	Vol. % Porosity
Al-Cu	As-cast	--	--	0.6
Al-Cu	As-cast	23 µm	13.9	7.9
Al-Cu	As-rheocast	23 µm	10.9	0.8

Density measurement

The results of density measurements conducted on the as-processed unreinforced, conventionally cast and rheocast composite specimens revealed density values of 2.71, 2.57 and 2.75 g cm $^{-3}$ respectively. The volume percent of the porosity computed using the experimentally determined density values and the results of acid dissolution tests are shown in Table 1.

Aging Studies

The results of aging studies conducted on the unreinforced, conventionally cast and rheocast composite samples are shown in Figure 1. The results exhibit the presence of a well defined hardness peak at 9 hrs for the unreinforced and as-cast composite samples and at 6 hrs for the rheocast composite samples. The results also reveal that the maximum peak hardness is achieved in the rheocast composite samples followed by the conventionally cast composite samples and lastly, the unreinforced samples.

Microstructural Characterization

Scanning electron microscopy conducted on unetched and etched unreinforced Al-Cu samples revealed the presence of: partly dendritic and partly equiaxed matrix microstructure, minimal amount of micrometer sized porosity and interdendritically located Cu rich intermetallic phase. A representative micrograph taken from the unreinforced sample is shown in Figure 2.

The results of scanning electron microscopy conducted on conventionally cast Al-Cu/SiC samples revealed a partly dendritic and partly equiaxed matrix microstructure. The interdendritic/intercellular regions were found to be frequently associated with the presence of Cu rich phases (see Figure 3). The metallic matrix also revealed the presence of porosity predominantly associated with the individual SiC particulates at the angular locations and with SiC clusters. The distribution of SiC particulates in the cast composite samples can be assessed from Figure 4. Predominantly, SiC particulates were present in the form of small clusters preferentially located at the grain boundaries. The interfacial integrity between SiC particulates and Al-Cu matrix was found to be poor and in some cases partially debonded interface was observed in the cast composite samples. In addition, the interface formed between the SiC particulates and Al-Cu matrix also revealed the presence of secondary phases. EDX analyses carried out at various locations in the Al-Cu/SiC interfacial locations revealed the enrichment of Cu. The results of EDX point analyses are graphically represented in Figure 5.

Finally the results of microstructural characterization studies carried out on the rheocast samples revealed the presence of: partially columnar and partially equiaxed matrix microstructure, minimal amount of porosity, Cu-rich phases at the interdendritic/intercellular regions, relatively more uniform distribution of SiC particulates and good interfacial integrity between SiC particulates and the metallic matrix. A representative SEM micrograph showing the microstructural aspects of the rheocast composite samples is shown in Figure 6. In addition, the results of EDX analyses revealed the copper content to decrease with increasing distance from the interface (see Figure 5).

X-Ray Diffraction Studies

The X-ray diffraction results corresponding to unreinforced, conventionally cast and rheocast composite samples were analyzed. The lattice spacings (d) obtained were compared with that of the pure aluminum, α-SiC, Al_2Cu, α-Al_2O_3, Al-Si-C phases and Al_4C_3. The results of the phase analyses are shown in Table 2.

Table 2. Results of X-ray diffractometry studies.

Material	Processing	Condition	Identified Phases
Al-Cu	Cast	Peak aged	Al
Al-Cu/SiC	Cast	Peak aged	Al, α-SiC
Al-Cu/SiC	Rheocast	Peak aged	Al, α-SiC, α-Al_2O_3

Mechanical Behavior

The results of ambient temperature testing on the unreinforced, conventionally cast and rheocast composite samples, aged to peak hardness, are summarized in Table 3. The results in Table 3 reveal that 0.2% yield stress (0.2 % YS), ultimate tensile strength (UTS) and ductility of the rheocast specimens are superior when compared to those of the unreinforced and conventionally cast composite samples. The mechanical properties of the conventionally cast composites, on the contrary, were found to be inferior even when compared to the unreinforced material.

Table 3. Results of room temperature mechanical properties.

Material	Processing	Condition	0.2 % YS (MPa)	UTS (MPa)	Ductility (%)
Al-Cu	Cast	Peak aged	93.8 ± 12.1	139.8 ± 27.1	5.2 ± 4.9
Al-Cu/SiC	Cast	Peak aged	58.1 ± 21.3	61.0 ± 21.6	1.2 ± 0.3
Al-Cu/SiC	Rheocast	Peak aged	126.1 ± 23.3	216.0 ± 8.4	6.3 ± 2.3

Discussion

Microstructure

The microstructure of the unreinforced samples, conventionally cast composite samples and rheocast composite samples revealed three common salient features:

a) presence of partially columnar and partially equiaxed matrix microstructure,
b) presence of porosity, and
c) presence of interdendritic Cu-rich phase.

The partially columnar and partially equiaxed structure, commonly referred as "ingot" type of structure [7] indicates that the remaining liquid temperature after the onset of solidification from the mold wall remained above the nucleation temperature. The underlying principles behind the development of "ingot" type of structure are well established and can be found elsewhere [7].

Another important microstructural feature observed in case of unreinforced and reinforced samples investigated in the present study was the presence of porosity. The formation of microporosity in the unreinforced and reinforced samples under the experimental conditions used in the present study was inevitable primarily as a result of columnar-equiaxed type of solidification structure observed in the present study. The mechanisms associated with the formation of porosity during solidification of materials exhibiting columnar-equiaxed type of structure have been previously established and can be found elsewhere [7]. Regarding the amount of porosity, the results of the present study revealed that the volume percent of porosity was less than 1% in case of unreinforced and rheocast composite samples when compared to about 7.9 % determined for the conventionally cast composite samples. The higher volume percent of porosity observed in the cast composite samples can be attributed to the physical properties of the metallic material containing suspended ceramic particulates and the solidification associated particulates' distribution in the metallic matrix. The results of microstructural characterization conducted on the conventionally cast composite specimens clearly indicated the presence of metal free zones at the sharp corners of the SiC particulates and within the SiC clusters (see Figure 4). The development of metal free zones at sharp corners of SiC particulates can primarily be attributed to the inability of the high viscosity metallic alloy to negotiate sharp corners while the presence of metal free zones within SiC clusters can be attributed to the inability of liquid metallic alloy used in the present study to infiltrate the micrometer sized crevices in the inefficiently packed SiC particulates' clusters formed ahead of moving solidification front during conventional casting.

The presence of interdendritic/intercellular Cu-rich phase observed in case of unreinforced, conventionally cast and rheocast composite samples can be attributed to the sluggish solidification front velocity achieved during primary processing of materials, rejection of Cu ahead of the moving liquid-solid interface and subsequent solidification when the temperature of the remaining liquid reached eutectic temperature [4, 7, 8]. The discontinuous nature of Cu-rich phase along the grain boundaries / interdendritic regions may be attributed to the partial dissolution during solutionizing step of the T6 heat treatment employed in the present study.

Amount and Distribution of SiC particulates

In the present study, 13.9 and 10.9 weight percent of SiC particulates were successfully incorporated in Al-2 wt.% Cu metallic matrix using conventional casting and rheocasting techniques respectively. The successful incorporation of SiC particulates in the limits exceeding 10 weight percent using these techniques can be attributed to the preheating of SiC particulates to 900 °C prior to the addition in the liquid metallic melt. Preheating of SiC particulates has been shown to assist in: i) removing surface impurities, ii) desorption of gases, and iii) altering the surface composition due to the formation of thin oxide layer (SiO_2) on the surface [9]. The ability of the oxide layer to improve the wettability of SiC particulates by alloy melt has previously been suggested by other investigators [10, 11].

Regarding the distribution of SiC particulates, following comments are in order. The preferential location of SiC particulates in case of cast composite samples is consistent with the results of other investigators and was attributed to the sluggish solidification front velocity commonly associated with the casting route[4, 7]. In the case of rheocast composite samples, the improved distribution of SiC particulates can be attributed to the enhanced mechanical interaction between SiC particulates and the broken dendrite fragments resulting from the stirring of the partial-liquid phase melt. The presence of dendrite fragments thus assist in better dispersion of reinforcing particulates and prevent settling of SiC particulates [12, 13].

Interfacial Characteristics

The results of microstructural characterization studies conducted on the particulate/matrix interfacial region revealed a high concentration of Cu in both conventionally cast and rheocast composite samples. This phenomenon can primarily be attributed to the presence of enhanced dislocation density in the interfacial region. The enhanced dislocation density results due to the difference in coefficient of thermal expansion between SiC particulates and the aluminum matrix [14] and promotes the dislocation-assisted diffusion of the alloying elements from the adjacent dislocation lean areas of the matrix. In related studies conducted on spray deposited aluminum based metal matrix composites, Gupta et al. [5] reported a similar enrichment of the main alloying element Cu in the SiC particulate/matrix interfacial region. The present experimental findings thus suggest that the interfacial segregation of the alloying elements is primarily governed by the physical properties of the metallic and ceramic components of the metal matrix composites and is independent of the type of processing technique employed to synthesize the composites.

Another important microstructural characteristic associated with the Al-Cu/SiC interfacial region in the case of conventionally cast composite samples was the presence of secondary phases (see Figure 3). These secondary phases were found to be located at and in near vicinity of SiC particulates demonstrating the heterogeneous nucleation capability of the interfacial region as a whole. The present experimental results thus indicate the heterogeneous nucleation capability of the SiC particulates and in addition the capability of the dislocations defect structure in the interfacial region in providing preferential sites for heterogeneous nucleation when compared to the bulk matrix. Further work is continuing in order to identify the composition and structural aspects of these secondary phases.

Aging Studies

The results of this study reveal that the aging kinetics remained the same for unreinforced and conventionally cast composite materials while it was accelerated in case of rheocast composite samples. The similar aging kinetics exhibited by the unreinforced and conventionally cast composite samples are consistent with the findings and observations of other investigators [5, 15, 16]. Chawla et al. [15] and Salvo et al. [16], for example, showed a negligible difference in aging kinetics of unreinforced and reinforced materials aged at relatively low temperatures (~ 150 °C). The present results are also supported by the findings of Gupta et al. [5] who reported the similar aging kinetics exhibited by spray processed Al-Cu, Al-Cu/SiC and Al-Cu/Al$_2$O$_3$ materials aged at 163 °C. These results are however in contradiction with the accelerated aging kinetics results obtained on reinforced aluminum based composite materials by other investigators [17-19].

The accelerated aging kinetics observed in case of rheocast composite samples aged at an identical aging temperature and containing less weight percent of SiC particulates when compared with conventionally cast composite samples is strongly indicative of the matrix microstructural variation brought about by the rheocasting technique. The fragmentation of dendrites during partial-liquid phase stirring [13] and the ensuing constrained growth during subsequent solidification appears to be instrumental in effecting the microstructural variation and the associated aging kinetics of the rheocast samples. Further work is continuing in this area.

Mechanical Behavior

The results of the present study revealed inferior strength (0.2 % YS and UTS) and ductility of the conventionally cast composite samples when compared to the unreinforced samples. This degradation in mechanical properties exhibited by the conventionally cast composite samples can be attributed to the coupled influence of :

a) presence of 7.9 volume percent porosity (see Table 1),
b) non uniform distribution of SiC particulates (see Figure 4) and
c) poor interfacial integrity between SiC particulates and Al-Cu matrix.

The porosity associated reduction in strength has been previously established by other investigators for steels, copper and aluminum based alloys and will not be reiterated here [20, 21].

The degradation in mechanical properties as a result of non uniform distribution of SiC particulates can be attributed to the tendency of early crack nucleation in the matrix at the clusters or agglomeration sites [5].

Another important feature that might have contributed towards the degradation in the strength of the conventionally cast composite material is the poor interfacial integrity observed between Al-Cu matrix and SiC particulates. The poor interfacial integrity prevents the effective load transfer across Al-Cu/SiC interface thus reducing the role of SiC particulates as load carriers in the metallic matrix. In analogous studies [5], it has been indicated that a strong interfacial bond contributes effectively towards the enhancement of strength of the composite materials.

Finally the results of the mechanical properties characterization revealed a significant improvement in the strength and ductility of the rheocast composite samples when compared to the conventionally cast unreinforced and composite samples. For example, the results shown in Table 3 indicate an increase in UTS and ductility of the rheocast samples by 1.5 and 1.2 times respectively when compared to the unreinforced samples and 3.5 and 5.3 times respectively when compared to the conventionally cast composite samples. The superior tensile properties thus exhibited by the rheocast samples when compared to either unreinforced samples or conventionally cast composite samples are attributed to the processing-associated improved microstructural uniformity, low volume fraction of porosity, relatively more uniform distribution of SiC particulates (minimum stress concentration sites) and good interfacial integrity between the ceramic particulates and Al-Cu metallic matrix.

Conclusions

The primary conclusions that may be derived from this work are as follows:

1. Aluminum based metal matrix composites containing upto 10.9 weight percent of SiC particulates can be successfully synthesized by rheocasting route used in the present study.
2. A low volume percent (0.8) of porosity observed in the rheocast samples when compared to the conventionally cast composite samples (vol. % = 7.9) is indicative of the potential of the rheocasting technique to make near net shape products.
3. The increase in strength and ductility of the rheocast composite samples can be attributed to the improved microstructural homogeneity and superior interfacial integrity between SiC particulates and Al-Cu matrix when compared to the conventionally cast material.

Acknowledgments

MG and MOL would like to thank NUS (grant # RP 3940619) for financial support during the course of this investigation. In addition, the authors would like to thank Mr Thomas Tan and Mr Tung Siew Kong (National University of Singapore, Singapore) for their valuable assistance and for many useful discussions.

References

1. P.S. Gillman, J. Met., 43 (8), 7 (1991).
2. A.L. Geiger and J.A. Walker, J. Met., 43 (8), 8 (1991).
3. I.A. Ibrahim, F.A. Mohamed and E.J. Lavernia, J. Mat. Sci., 26, 1137 (1991).
4. M. Gupta, C. Lane and E.J. Lavernia, Scr. Metall. et Mater., 26, 825 (1992).
5. M. Gupta, T.S. Srivatsan, F.A. Mohamed and E.J. Lavernia, J. Mat. Sci., 28, 2245 (1993).
6. Powder Diffraction File, International Center for Diffraction Data, 1601 Park Lane, Swarthmore, PA, USA, 1991.
7. B. Chalmers, Principles of Solidification, pp. 253-297 John Wiley and Sons Inc., New York, USA, 1964.
8. Metallography and Microstructures, Metals Handbook, 9th edition, Vol. 9, p. 632, ASM, Metals Park, Ohio, USA, 1986.
9. C.M. Milliere and M. Suery, Mat. Sci. and Tech., 4, 41 (1988).
10. H. Ribes, R. Dasilva, M. Suery and T. Bretheau, Mat. Sci. Tech., 6, 621 (1990).
11. L.N. Thanh and M. Suery, Scr. Met., 25, 2781 (1991).
12. F.A. Girot, L. Albingre, J.M. Quenisset and R. Naslain, J. Met., 18 (1987).
13. R. Mehrabian, R.G. Riek and M.C. Flemings, Metall. Trans., 5, 1899 (1974).
14. R.J. Arsenault and N. Shi, Mat. Sci. and Eng., 175 (1981).
15. K.K. Chawla, A.H. Esmaeili, A.K. Datye and A.K. Vasudevan, presented at TMS/ASM fall meeting, Cincinnati. Ohio, USA, Oct. 1991.
16. L. Salvo, M. Suery and F. Decomps, in " Fabrication of Particulates Reinforced Composites", edited by J. Masounave and F.G. Hamel, (ASM International, Materials Park, Ohio, 1990), p. 139.
17. T. Christman, A. Needleman, S. Nutt and S. Suresh, Mat. Sci. and Eng., 107A, 49 (1989).
18. T.G. Nieh and R.F. Karlak, Scr. Met., 18, 25 (1984).
19. J.M. Papazian, Metall. Trans., 19A, 2945 (1988).
20. G.F. Bocchini, The Int. J. of Powder Metallurgy, 22 (3), 185 (1986).
21. R.D. Payne, A.L. Moran and R.C. Cammarata, Scr. Metall. et Mater., 29, 907 (1993).

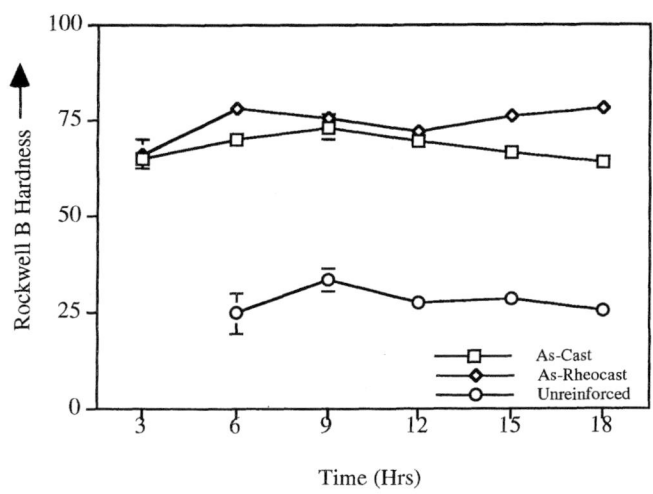

Figure 1 Graphical representation of aging studies conducted on conventionally cast Al-Cu and Al-Cu/SiC samples and rheocast Al-Cu/SiC samples.

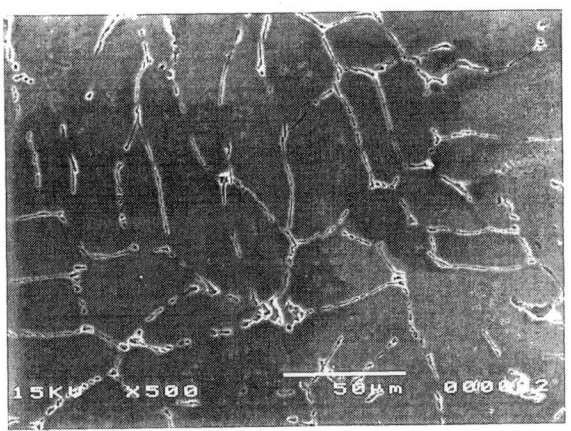

Figure 2 Representative SEM micrograph showing the microstructural features observed in conventionally cast Al-Cu samples.

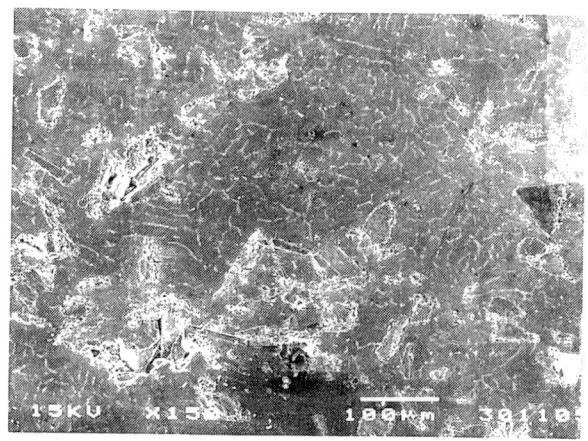

Figure 3 Representative SEM micrograph showing the presence of Cu-rich phases in the interdendritic/intercellular regions in conventionally cast Al-Cu/SiC samples.

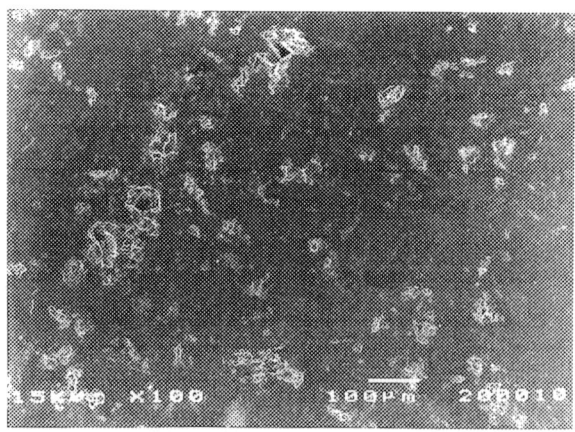

Figure 4 Representative SEM micrograph showing the distribution of SiC particulates in conventionally cast Al-Cu/SiC samples.

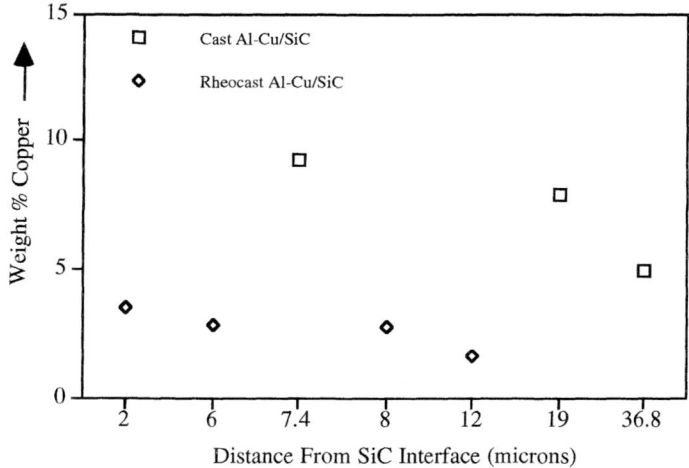

Figure 5 Graphical representation of the segregation pattern of Cu observed at Al-Cu/SiC interfacial regions in case of conventionally cast and rheocast Al-Cu/SiC samples.

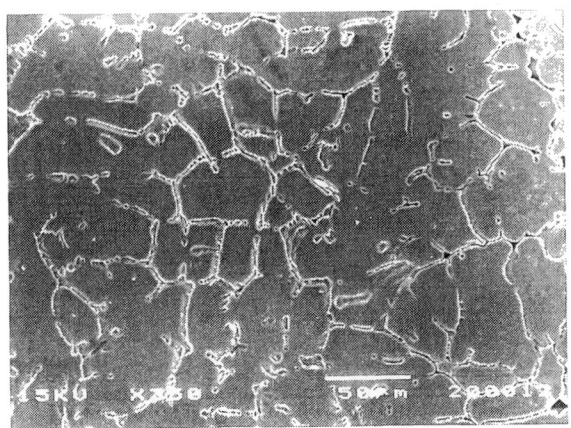

Figure 6 Representative SEM micrograph showing the microstructural features of the rheocast Al-Cu samples.

HIGH TEMPERATURE DEFORMATION OF MAGNESIUM AND ALUMINUM ALLOY COMPOSITES REINFORCED BY SiC WHISKERS

Makoto SUGAMATA

College of Industrial Technology, Nihon University
Narashino, Chiba 275, Japan

ABSTRACT

SiC whisker reinforced AZ91 magnesium and 2324 aluminum alloy composites were fabricated by pressure infiltration of the alloy melt into the whisker preform and subsequent hot extrusion of the composite ingots after over-aging heat treatment. Tensile strength and elongation at elevated temperatures of as-extruded composites were measured by tensile test at the initial strain rates ranging from $5 \times 10^{-3} s^{-1}$ to $5 \times 10^{-1} s^{-1}$. Metallographic structures were studied on as-extruded composites by optical microscope and TEM. Distribution of SiC whiskers and morphology of voids near the fracture surface were observed by SEM.

A fine subgrain structure of less than $1 \mu m$ in diameter was observed for the 2324 aluminum alloy matrix of the as-extruded composites. Tensile strength of the extruded composites decreases more markedly with the test temperature and becomes even lower than that of the matrix alloy above 623K for SiCw/AZ91 composites, and 673K for SiCw/2324 composites. At high temperatures above the solidus temperature of the the matrix alloy, tensile strength of the both extruded composites decreases to around 1MPa. The strain rate sensitivity(m-value) of extruded composites is generally higher than that of the matrix alloy at high temperatures. However, correlation between the m-value and elongation is lower for the both extruded composites. The Vf20% composite of SiCw/2324 extruded with the ratio of 1:49 shows superplastic elongation as high as 520% at 793K at the initial strain rate of $5 \times 10^{-2} s^{-1}$, and elongation higher than 200% at other strain rates. The highest elongation of SiCw/AZ91 composites extruded with the ratio of 1:25 is about 120% at 673K at an initial strain rate of $5 \times 10^{-1} s^{-1}$. No appreciable necking was observed during deformation under these tensile conditions of the both composites. Therefore, for the extruded composites, both a fine grained structure with high thermal stability and partial melting are necessary to cause superplasticity. It is postulated that SiCw/matrix interface sliding occurs in superplastic deformation along with grain boundary sliding.

INTRODUCTION

Superplastic behavior has been reported in several ceramics whisker or particle reinforced aluminum alloy composites at high temperatures and high strain rates[1]-[3]. The fine grain microstructure of the matrix was observed in the composites which were fabricated by mixing reinforcement with metallic powder and subsequent hot extrusion of mixture with high reduction. It becomes clear that the fine grain microstructure is required for composites to show superplasticity, but fine grain structures are not sufficient for composites to show superplasticity[4]. Grain boundary sliding and sliding at the interface between reinforcement and matrix are considered to cause superplasticity. Solidus temperature at the interface falls because alloying elements tend to segregate at the interface. So, the interface sliding takes place at lower temperature than melting temperature of matrix[2].

High pressure casting procedure is advantageous for industrially fabricating metal matrix composites. It is expected that superplastic behavior occurs on those composites if microstructures are

Inorganic Matrix Composites
Edited by M. K. Surappa
The Minerals, Metals & Materials Society, 1996

properly prepared by high temperature deformation. However, very few researches concerning superplasticity of the composites via casting route have been done[5]. In this study, SiC whisker reinforced AZ91 magnesium and 2324 aluminum alloy composites were fabricated by pressure infiltration of alloy melt and hot extrusion after over-aging treatment in order to refine grain structure of extruded composites. The purpose of this paper is to make clear superplastic behavior of those extruded composites by examining the tensile properties of extruded composites via high pressure casting route.

EXPERIMENTAL PROCEDURE

Silicon carbide whiskers of β-cubic structure were used for the reinforcement of AZ91 magnesium and 2324 aluminum alloys. Cylindrical preforms of ϕ 35 × 90mm with two different packing densities, 13% and 20%, were used for pressure infiltration of the alloy melt to fabricate cast composites of two different volume fractions of the reinforcement. SiC whisker preforms used in this work were supplied by Tokai Carbon Co., Ltd. The supplier's reference data of SiC whiskers are shown in Table 1. The fabrication processes of the composites in this work are schematically shown in Fig.1. Each alloy was melted and infiltrated into a SiC whisker preform under pressure of 100MPa in air or SF_6+CO_2 mixture gas. The SiC preforms were preheated to 1173K for 1.2ks prior to pressure infiltration. The cast composites were over-aged after solutionizing for 7.2ks prior to hot-extrusion. It was reported that distribution of relatively coarse precipitates produced by over-aging had an effect of grain refining during dynamic recrystalization[6]. Hot-extrusion was done at 633K for AZ91 alloy and at 673K for 2324 alloy under estimated strain rate of $8.8 \times 10^{-3} s^{-1}$ into rods of 5mm, 7mm and 10mm in diameter. Extruded rods without SiC whiskers were also prepared by same fabrication processes of composites.

Constant velocity tension tests were carried out in the range between $5 \times 10^{-3} s^{-1}$ and $5 \times 10^{-1} s^{-1}$ at temperatures between 623K and 793K. The solidus temperature is 694K and 775K, and liquidus temperature is 743K and 911K for AZ91 and 2324 alloy matrix respectively. Distribution and alignment of

Table 1 Physical properties of SiCw

Crystal	Cubic (β)
Diameter	0.1~1.0 μm
Length	30~100 μm
Aspect Ratio	50~200
Density	3.19 Mg/m^3
Tensile Strength	3~14 GPa
Elastic Modulus	400~700 GPa

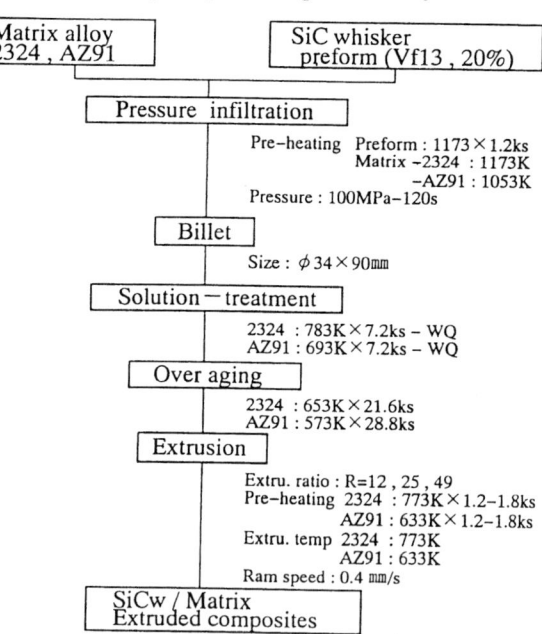

Fig.1 Fabrication process of SiCw/AZ91 and SiCw/2324 composites.

SiC whiskers in extruded composites, fracture surface and void formation of tensile tested specimens were examined by SEM. Microstructural observation in the matrix of as-extruded composites was done by means of TEM. The TEM specimens of 2324 alloy matrix composites were thinned electorolytically in a 30%HNO_3 -70%CH_3OH solution at approximately 253K.

RESULT AND DISCUSSION

AZ91 magnesium alloy matrix composites

Microstructure of extruded composites

Fig.2 shows distribution and morphology of SiC whiskers in cast composites and in extruded composites. In cast composites, longer SiC whiskers are distributed in three dimensionally random directions. However, short SiC whiskers broken during extrusion are aligned in the extrusion direction in extruded composites.

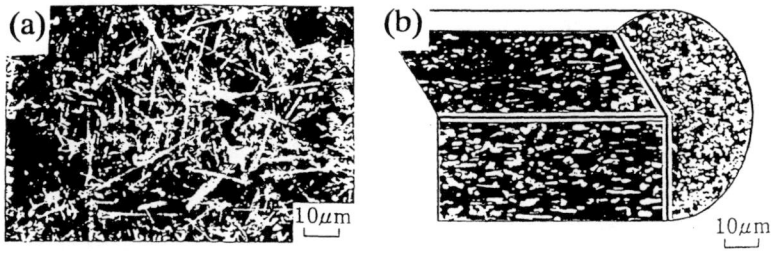

Fig.2 SEM micrographs of cast composites(a) and extruded composites(b)
(SiCw/AZ91, Vf20%, R=49)

Mechanical properties of extruded materials.

Fig.3 shows tensile strength and elongation of matrix alloy and composites extruded at extrusion ratio of 1:49 against testing temperatures at lower initial strain rate. The composites show higher tensile strength than matrix at room temperature, but at testing temperatures above

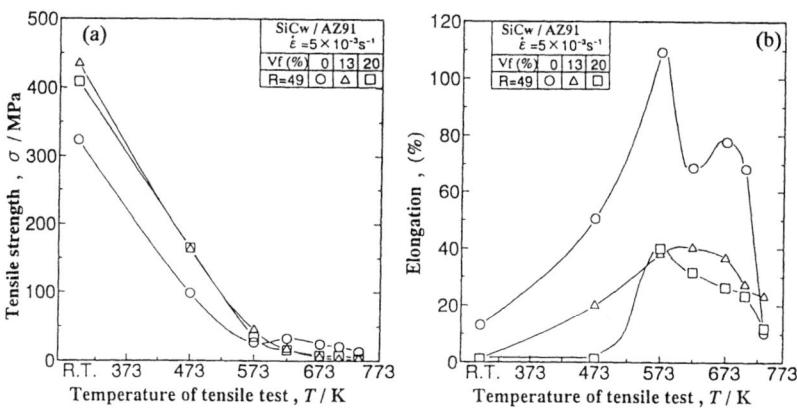

Fig.3 Temperature dependence of tensile strength(a) and elongation(b) of AZ91 alloy and SiCw/AZ91 extruded composites.

623K, composites show lower strength than the matrix. It was reported that the fine grained microstructures of extruded composites led to decreasing of the strength at high temperatures due to increased contribution of grain boundary sliding[6]. The elongation is 12% for matrix and 1% for composites at room temperature. The matrix alloy shows the highest elongation of about 110% at 573K, and fairly high elongation below 673K. At 733K, elongation of matrix alloy decreases due to partial melting. The composites show higher elongation than that of matrix alloy at 733K.

Effect of strain rate and temperature on the deformation behavior

Fig.4 shows strain rate dependence of peak flow stress of the matrix alloy and the composites extruded at extrusion ratio of 1:25 at different testing temperatures. The strain rate sensitivity(m-value) were obtained by measuring inclination in this figure for each material. The m-value of the

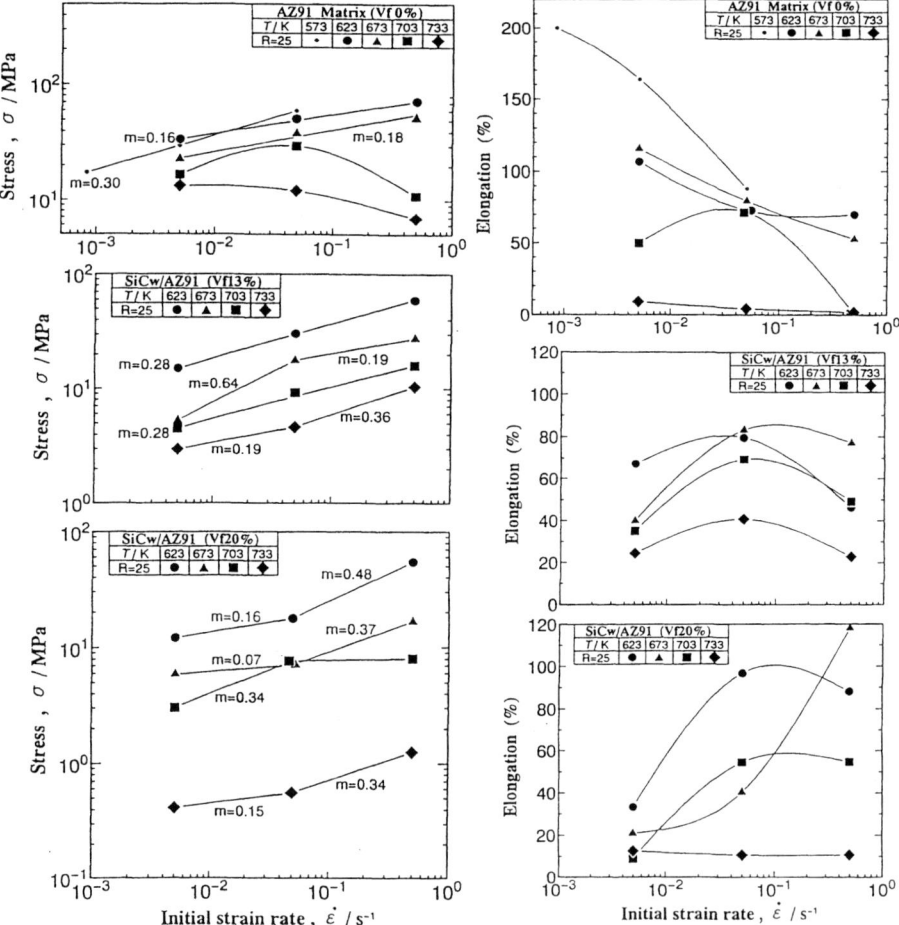

Fig.4 Strain rate dependence of peak flow stress of AZ91 alloy and SiCw/AZ91 composites at various temperatures.

Fig.5 Strain rate dependence of elongation of AZ91 alloy and SiCw/AZ91 composites at various temperatures.

matrix alloy is approximately 0.30 at 573K, but decreases with increase of testing temperature. At higher testing temperatures, the flow stress of the matrix alloy tends to drop at high strain rate. The Vf13% composites show m-value between 0.19 and 0.64. Composites tend to show increasing m-values with increasing strain rate at higher temperature. The strain rate dependence of elongation is shown in Fig.5. For the matrix alloy, the highest elongation of about 200% is obtained at 573K at comparatively lower strain rate, and elongation decreases along with testing temperature and strain rate. The Vf13% composites show higher elongation at an initial strain rate of $5 \times 10^{-2} s^{-1}$, and the highest elongation of 80% is obtained at 673K. The Vf20% composites tend to show increasing elongation with increasing strain rate. For the Vf20% composites, the highest elongation of 120% is obtained at the highest strain rate at 673K. Increase of elongation was not observed on the composites extruded at higher reduction ratio.

Fracture surface of tensile tested specimens.

Fig.6 shows fracture surfaces of Vf20% composites extruded at reduction ratio of 1:25 at an initial strain rate of $5 \times 10^{-2} s^{-1}$ at three different testing temperatures. The elongations were 120% at 673K, 60% at 703K and 10% at 733K. Fine granular fracture surface can be observed on all specimens. These fine granular surfaces suggest occurrence of grain boundary sliding and grain boundary fracture at higher temperature. Interfacial sliding between the matrix and whisker in addition to grain boundary sliding is considered to contribute to increase elongation in the composites. For SiCw/AZ91 composites, it was considered that oxidation of the matrix prevented occurrence of interfacial sliding between the matrix and whisker at partial melting temperatures.

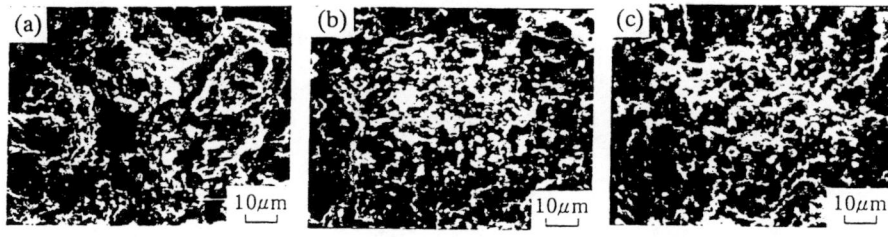

Fig.6 Fracture surfaces of SiCw/AZ91(Vf20%, R=25) composites after tensile test at an initial strain rate of $5 \times 10^{-2} s^{-1}$. (a):673K, (b):703K, (c):733K

2324 aluminum alloy matrix composites

Microstructure of extruded composites

Shorter SiC whiskers broken during extrusion were oriented in the extrusion direction in extruded composites. A TEM micrograph of Vf13% composites extruded at a ratio of 1:49 is shown in Fig.7. Very fine subgrains of less than 1 μm in diameter are observed in the matrix. These fine subgrain microstructures were formed by hot extrusion; the subgrains become finer with increasing extrusion ratio and whisker volume fraction.

Fig.7 TEM micrograph of as-extruded SiCw/2324 composites(Vf13%, R=49) in the longitudinal section.

Mechanical properties of extruded materials.

Tensile strength of as-extruded materials is plotted on a logarithmic scale against test temperatures in Fig.8. It is noted that the composites show higher tensile strength than matrix at room temperature, but strength decrease of composites with increasing test temperature is larger than that of the matrix alloy. At testing temperature of 773K, composites show strength lower than 1MPa.

Fig.9 shows nominal stress-strain curves of the matrix alloy and Vf13% composites at various testing temperatures at an initial strain rate of $5 \times 10^{-2} s^{-1}$. At 693 and 743K, the flow stress of the matrix alloy shows rapid decrease after reaching its maximum, while the flow stress of the composites shows more gradual decrease. At 793K above solidus temperature of the matrix alloy, fracture of the matrix alloy occurs at lower strain, whereas the composites deform under very low flow stress up to very high strain. Similar deformation behavior was observed on the Vf20% composites.

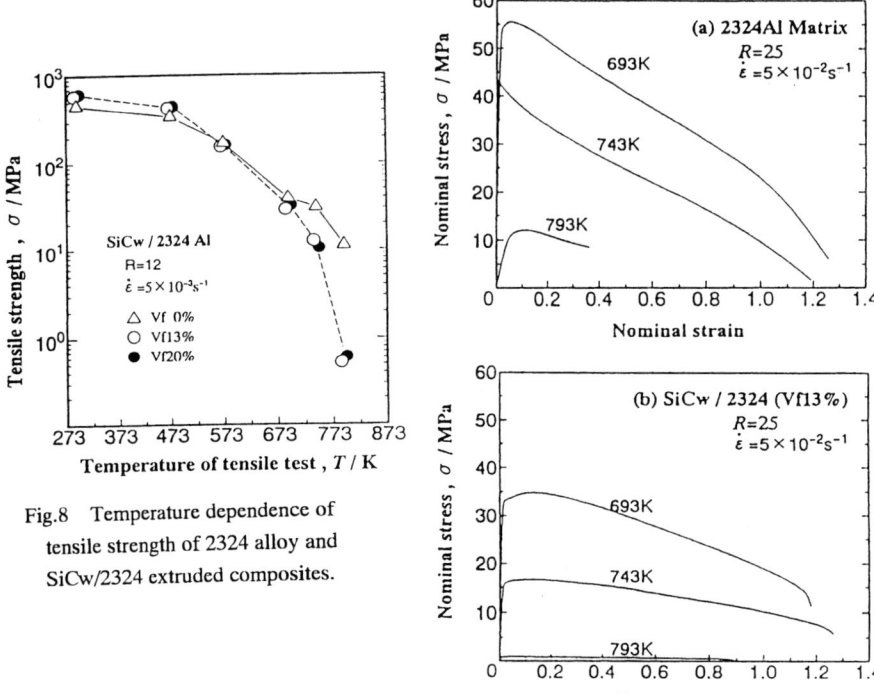

Fig.8　Temperature dependence of tensile strength of 2324 alloy and SiCw/2324 extruded composites.

Fig.9　Typical nominal stress-strain curves for 2324 alloy and SiCw/2324 (Vf13%) composites at three different test temperatures.

Effect of strain rate and temperature on the deformation behavior

Fig.10 shows strain rate dependence of peak flow stress of the matrix alloy and composites extruded at various extrusion ratio at three different testing temperatures. The m-value of the matrix alloy is approximately 0.16 at both 693 and 743K. The Vf13% composites show m-value between

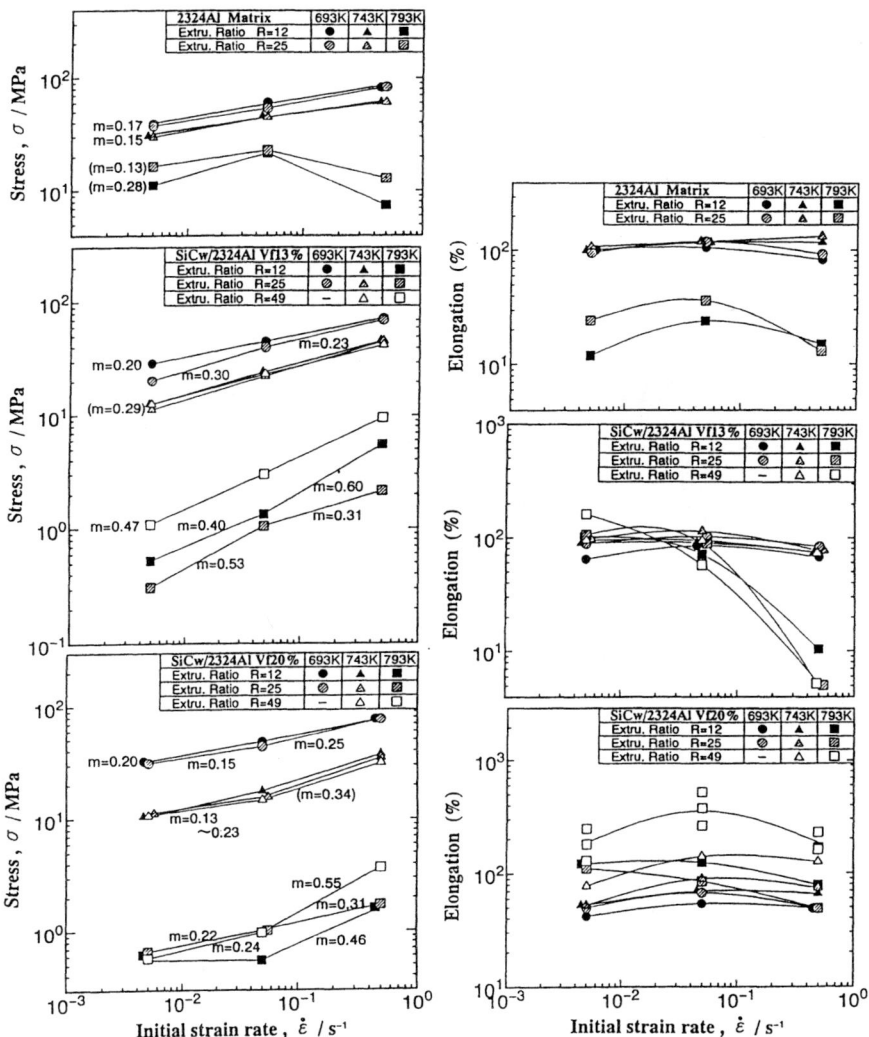

Fig.10 Strain rate dependence of peak flow stress of 2324 alloy and SiCw/2324 composites at various temperatures.

Fig.11 Strain rate dependence of elongation of 2324 alloy and SiCw/2324 composites at various temperatures.

0.2 and 0.6, which increases with testing temperatures. The m-value of Vf20% composites is higher at high strain rate. It is clear that composites show higher m-values at 793K and at high strain rate. The strain rate dependence of elongation of those materials is shown in Fig.11. Elongation of the matrix alloy is as high as 100% at both 693 and 743K, but decreases to around 40% at 793K. At 743K, the elongation of the matrix alloy increases as the strain rate increases. Elongation of Vf13% composites is higher than that of the matrix alloy at 793K at initial strain rate of $5 \times 10^{-3}s^{-1}$ and $5 \times 10^{-2}s^{-1}$, and the Vf13% composites (R=49) shows the highest elongation of about 200% at lower strain rate. Elongation of Vf20% composites is lower than that of Vf13% composites at 693K at any strain rates. At 743K, however, Vf20% composites (R=49) show superplastic elongation as high as 520% at an initial strain rate of $5 \times 10^{-2}s^{-1}$, and large elongation of more than 200% at any other strain rates.

Tensile test specimens of the matrix alloy deformed to failure at 743K(b) and 793K(c) are shown in Fig.12. Plastic strain was concentrated at near fracture point for the specimen(b) showing 130% elongation. Fracture occurred without necking for a specimen(c) tested at 793K due to partial melting.

Fig.13 shows specimens of Vf20% composites(R=49) tested at 793K with three different strain rates. All specimens deformed without necking. The specimen(c) tested at an initial strain rate of $5 \times 10^{-2}s^{-1}$ shows superplastic elongation of 520%.

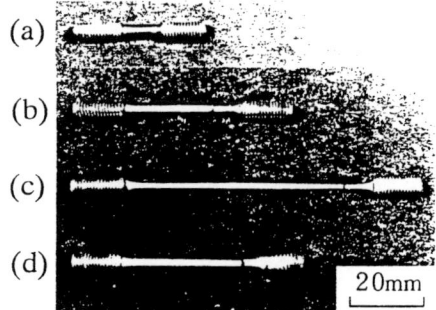

Fig.12 Tensile test specimen of 2324 alloy (R=25).(a)prior to test,(b)743K and (c)793K at an initial strain rate of $5 \times 10^{-2}s^{-1}$

Fig.13 Tensile test specimens of SiCw/2324 (Vf20%, R=49) composites. (a)prior to test, after tensile test at 793K at(b)$5 \times 10^{-3}s^{-1}$,(c)$5 \times 10^{-2}s^{-1}$ and (d)$5 \times 10^{-1}s^{-1}$.

Microstructures of tensile tested specimens

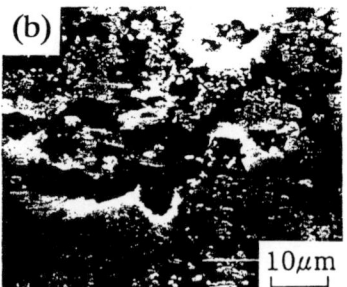

Fig.14 Fracture surfaces (a) and longitudinal section near fracture surface (b) of SiCw/2324 (Vf20%, R=49) composites after tensile test at 793K at an initial strain rate of $5 \times 10^{-2}s^{-1}$.

Fig.14 shows fracture surface and longitudinal section near the fracture surface of Vf20% composites(R=49) after tensile test at 793K at an initial strain rate of $5 \times 10^{-2} s^{-1}$. Smoother fracture surface suggests occurrence of partial melting of the matrix at grain boundary. Unevenness of the fracture surface is finer for composites than that for the matrix alloy. Hence, deformation of composites is in part caused by grain boundary sliding. It can be seen on Fig.14(b) that SiC whiskers distribute lessuniformly after tensile test than those in untested specimens.

SEM micrographs are shown in the longitudinal section near fracture surface of Vf20% composites(R=49) tested at 793K at different initial strain rates in Fig.15. The increase of strain rate tends to suppress the void growth in tensile direction. Fracture occurs by interconnection of voids during deformation. Hence, in composites the highest elongation was obtained at comparatively high strain rate. Interfacial sliding between the matrix and whisker in addition to grain boundary sliding is considered to contribute to increase of superplastic elongation in the composites. So, sufficient amount of grain boundaries and SiCw/matrix interface are required to cause superplasticity. Elongation of Vf13% composites tested at 793K is lower than that of Vf20% probably due to insufficient amount of grain boundaries and SiCw/matrix interface.

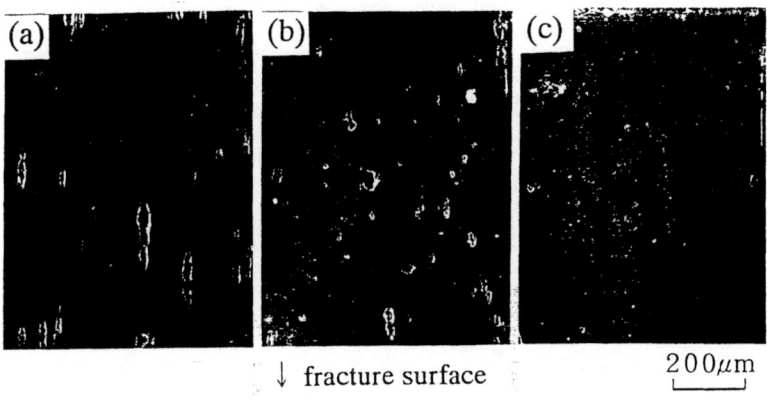

↓ fracture surface $200\mu m$

Fig.15 SEM micrographs of SiCw/2324 (Vf20%,R=49) composites in the longitudinal section near fractured surfaces after tensile test at 793K, at (a)$5 \times 10^{-3} s^{-1}$, (b)$5 \times 10^{-2} s^{-1}$ and (c)$5 \times 10^{-1} s^{-1}$.

SUMMARY

SiCw/AZ91 and SiCw/2324 composites were fabricated by high pressure casting and subsequent hot extrusion. Mechanical properties of their composites were measured at room and elevated temperatures. The results obtained from experiments are as follows:
(1) Short iC whisker broken by extrusion were oriented in the extrusion direction in the composites. A fine subgrained structure was observed for the matrix of as-extruded composites.
(2) The composites show higher tensile strength at room temperature, but lower strength at elevated temperature than the matrix alloy.
(3) The Vf20% composite of SiCw/2324 extruded with the ratio 1:49 shows superplastic elongation of as high as 520% at 793K at an initial strain rate of $5 \times 10^{-2} s^{-1}$. However, superplastic elongation is not observed on SiCw/AZ91 composites.
(4) Correlation between the m-value and elongation is lower for the composites.
(5)SiCw/matrix interface sliding occurs in superplastic deformation along with grain boundary sliding in the matrix.

REFERENCES

[1] A.H.Choksi, T.R.Bieler, T.G.Nieh, J.Wadsworth & A.K.Mukherjee : Superplasticity in Aerospace, TMS, (1988), 229.
[2] T.G.Nieh & J.Wadsworth : Superplasticity in Advanced Materials, (1991), 339, JSRS.
[3] M.Mabuchi, T.Imai & K.Kubo : Japan Inst.Light Metals,41(1991), 108.
[4] J.S.Kim, M.Sugamata & J.Kaneko : J.Japan Inst. Met., 55-9(1991), 994.
[5] H.Xiaoxu, L.Qing, C.K.Yao & Y.Mei: J. Mater. Sci.Lett., 10(1991), 964.
[6] J.A.Wert, N.E.Paton, C.H.Hamilton & M.W.Mahony : Metall. Trans. 12A(1981), 1267.
[7] J.S.Kim, J.Kaneko & M.Sugamata : J.Japan Inst. Met., 56-7(1992), 819.

TECHNOLOGY DEVELOPMENT OF CAST ALUMINIUM BASED COMPOSITES

B.C. Pai, R.M. Pillai and K.G. Satyanarayana
Regional Research Laboratory (CSIR)
Trivandrum 695 019, India

ABSTRACT

Realising the potentials of discontinuous reinforced aluminium alloy matrix composites (AMC) in engineering applications the work was initiated at the Regional Research Laboratory, Trivandrum (RRL-T) a decade back. The laboratory has chosen stir casting technique including both liquid metal processing and semi-solid slurry technique for the synthesis of AMC. Among the various dispersoid systems investigated such as coconut shell char, zircon, zirconia, titania, flyash, glass, graphite, short carbon fibres and silicon carbide, the last three systems were studied in detail with respect to synthesis, structure property correlation and product development. The graphite particle dispersed composite exhibited adhesive wear resistance and components demonstrated self lubrication properties. The hot extruded short carbon fibre composites showed improved properties for structural applications. The silicon carbide dispersed composites exhibited good wear resistance and mechanical properties for various engineering applications including automotive. The paper overviews the work carried out on AMC at RRL-T.

INTRODUCTION

The unique tailorability, higher temperature withstanding capability, improved mechanical properties and some special properties are some of the advantages realised with metal matrix composites (MMC) compared to monolithic base alloy. The work on discontinuous reinforced aluminium alloy matrix composite (AMC) was initiated at the Regional Research Laboratory, Trivandrum (RRL-T) in the late seventies. Stir casting, i.e. both liquid metal processing[1] and semi-solid slurry casting[2] techniques is the method used for synthesis of AMC. The initial work was on the feasibility of making AMC with different particulate dispersoids like coconut shell char[3], zircon[4], flyash[5], zirconia[6], titania[6,7], glass[5] and graphite[8]. Studies related to synthesis and structure property correlations were carried out. Realising the potentials of graphite dispersed AMC for self lubricated bodies, detailed studies were carried out[9] with this system such as surface treatment to the graphite for better interfacial bonding, mixing, additions of alloying elements to the matrix alloy for improved wetting, solidification studies under different casting conditions[10] etc. These understanding have enabled to produce composite of uniform and consistent properties[9]. Similar detailed investigations were also carried out with AMC containing short fibres[11] of carbon and particulates of silicon carbide (SiC_p)[12]. Efforts were also made for developing and evaluating AMC components in collaboration with the user industries. Some of them are now undergoing field trials. This paper overviews the AMC work carried out at RRL-T including the component development and evaluation.

SYSTEMS INVESTIGATED AND DETAILS

Table I lists the composite systems investigated in the laboratory and some specific observations made with them.

Coconut shell char[3]:

Upto about 30 vol % of coconut shell char could be incorporated in Al alloy matrix. The presence of particles in the matrix reduced the tensile properties. Above 5 vol % of the shell char in the matrix could impart adhesive wear resistance to the matrix. The composite made with this system can be used for mild wear resistant applications under low loads and pressures.

Zircon ($ZrSiO_4$):

Upto 30 wt % zircon could be introduced[4] in aluminium alloy matrix. The presence of zircon particles in aluminium increased its hardness as well as improved its abrasive wear resistance[13]. In addition to gravity diecasting, pressure diecasting studies were also carried out with this system. The self-locking water taps (Fig. 1) made out of 15 wt% zircon composites could replace gun metal taps[14]. Composite also takes good electroplating with Ni and Cr under controlled condition.

Zirconia (ZrO_2)[6]:

About 5 vol % zirconia particles (2-10 μm size range) could be introduced in Al alloy matrix. Hardness and tensile properties were improved by the additions. This system has the potential for high strength/high temperature material.

Titania (TiO_2):

The semi-solid slurry casting technique was efficient for dispersing fine size range (0.01 to 10 µm) titania particles in aluminium alloy matrix compared to liquid metal processing[7]. In wrought alloy matrix (7020) with 5 wt% TiO_2, both hot and cold rolled composites exhibited superior ultimate tensile strength, and % elongation at room temperature as well as at 523 K compared to base alloy properties prepared and tested under similar conditions. The cold formability of the composite was much superior to base alloy specially for making thin sheets[15]. The composite has shown good potential for better workability, and improvements in the properties.

Flyash:

About 30 vol % of the flyash could be introduced in aluminium alloy matrix. With increased incorporation of the flyash, the ultimate tensile strength, % elongation and density of the composite decreased. The composite could be hot forged or worked to improve the mechanical properties[5].

Glass powder:

The ground borosilicate glass particulates incorporated in AMC were getting fibrised during extrusion in the direction of extrusion. The tensile property of this extruded AMC was much superior to the base alloy. The composite also exhibited good corrosion resistance under ambient conditions compared to the base alloy[5].

Graphite powder:

Both liquid metal processing[8] and semi-solid slurry casting techniques[16] are used for synthesising AMC. By semi-solid slurry technique a master composite of about 50 vol % graphite can be prepared

(Fig. 2) which can be diluted to the required level by dissolving it in the base alloy. There was no difference between the composites prepared by direct liquid metal processing or by the dilution of the master composite in terms of microstructure (Fig. 3a & b) and properties. The gravity segragation of the graphite particles can take place if solidification conditions are not properly controlled. Generally, permanent metal moulds having higher heat transfer rates are preferred for better distribution of the graphite. Pressure die casting and squeeze castings gave better distribution of the particles in the castings. By resorting to centrifugal casting, one can get higher graphite concentrations in the inner periphery of a hollow cylinder which can be used for bearing applications[17].

The macroscopic distribution of the graphite particles in composite castings can easily be assessed by removing the surface layer of the AMC by machining or by filing. Eventhough the gas content in the composite is higher than the base alloy, no serious gas porosities are observed in the composites. The typical mechanical properties of the composites with 5 vol % graphite in different alloy systems made in the laboratory with different techniques are listed in Table II. Presence of about 5 vol % graphite reduced the UTS of the base alloy by about 10-15% and % elongation to less than 50% of the base alloy. The graphite addition reduced the coefficient of friction from 0.35 of the base alloy to 0.25 with 5 vol % graphite which reaches a constant value of 0.2 when graphite content in the matrix exceeds about 15 vol %. The adhesive wear resistance also increased with the increase in the graphite content[18].

Studies[19] have shown that the composite can be remelted 3-4 times without deteriorating the properties appreciably. Hence, the recycling of the composite is also possible.

Short carbon fibre:

Short or chopped carbon fibres of length 1, 3, and 6 mm were introduced in aluminium alloy matrix[11,16]. A special treatment has been devised for separating the carbon fibres from the tow of 3000/6000 fibres without affecting its mechanical properties[20]. The direct introduction of the carbon fibres in the molten aluminium resulted in severe reaction between the fibre and the matrix. Metal coating like Cu and Ni over the fibres could minimise the reaction[21]. However, these coating introduced high levels of alloying in base aluminium alloy which affected the properties. Semi-solid slurry casting route for dispersing carbon fibres in Al alloy matrix could give reaction free composite with random distribution of the carbon fibre[11,16]. By hot extrusion the fibres can be aligned in the direction of extrusion (Fig. 4). The typical mechanical properties of the base alloy and the composites are given in Table III. About 8% carbon fibre improves the UTS by about 10% and modulus by about 15%. The marginal improvements in the properties can only be attributed to the shorter length of the carbon fibres which is below the critical length. It has been observed that during processing fibres get cut and shortened below the critical length.

Silicon carbide particulates (SiC_p):

Abrasive grade silicon carbide particulates of average particle size (APS) 14, 23, 34 and 43 µm were dispersed in both cast and wrought aluminium alloy matrix using both liquid metal processing and semi-solid slurry casting technique[12]. The oxidation treatment given to the SiC_p prior to synthesis helped in having a layer of $MgAl_2O_4$ spinel at the particle matrix interface. This layer helps in reducing the severeity of the reaction between SiC_p and molten aluminium even with low Si contained alloys like 6061 and 2024 during liquid metal processing. No appreciable gravity segregation of SiC_p was observed in permanent mould cast ingots of 100 mm Ø and 300 mm height or in shaped castings. The microstructure (Fig. 5(a)-(b)) and mechanical properties (Table IV) of the SiC_p dispersed Al-7 Si-Mg alloy matrix composites prepared in the laboratory are comparable with those of Duralcan, USA composites[22] and Hydro Aluminium[23] Corporation reported in the literature[24]. The formation of $MgAl_2O_4$ spinel at the particle matrix interface consumes some Mg from the matrix. Similarly Si released from the oxidised SiC according to the reaction (1) also depletes Mg from the matrix to form Mg_2Si.

$$SiO_{2(s)} + 2 Mg_{(l)} \longrightarrow 2 MgO_{(s)} + Si_{(s)} \qquad (1)$$

These are likely to affect the strengthening of the composites during heat treatment. Hence, optimisation of the Mg content in the matrix is very important. It was also observed that the presence of $MgAl_2O_4$ spinel at the interface helps in reducing the reaction during the remelting of these composites[12] by suppressing Al_4C_3 formation.

COMPONENT DEVELOPMENT AND EVALUATION:

The component development exercise was initiated after the basic properties and limitations of the composites were well established. In the absence of clear specifications and reliable data on the components, the MMC components could only be developed and evaluated on trial basis in collaboration with the user. Hence, prototype components were developed and evaluated with zircon, graphite particles and SiC_p dispersed composites. Some of the evaluation work are still under progress. Realising the abrasion resistance properties of zircon, self locking water tapes (Fig. 1) were made by pressure diecasting in collaboration with the user[14]. The composite taps performed better than the alloy taps.

In graphite dispersed aluminium alloy composite system, composite contactors (Fig. 6) were made by pressure diecasting technique[25]. During composite diecasting slight change in the die design was found to be essential taking into consideration of the high viscosity and higher gas content of the composite melt. The composite contactors could replace phosphor bronze contactor with better performance and with about 50% cost. The annual requirement of such contactor was about 3000 Nos. only and weight of each contactor was about 10 grams. Hence the commercial production of these composite components was uneconomical.

Sleeve bearings up to 100 mm diameter were made with Al-5% graphite composites and were evaluated in place of Al-Sn bearing (SAE 770) alloys as well as Cu base bearing alloys. The composite could replace SAE 770 alloy directly on one to one basis without much changes in the design[17]. However, with Cu base bearing alloys i.e. gun metal and phosphor bronze it was always not possible to replace them by composites. This was mainly due to higher coefficient of thermal expansion (CTE) as well as lower compressive strength of matrix aluminium alloys compared to copper base alloys. By having higher percentage of graphite ($>$ 20 vol %) in the matrix CTE and coefficient of friction can be brought down[18] and range of applications can be extended.

The application of graphite dispersed aluminium matrix composites for automotive applications has also been considered[26]. The feasibility of making graphite dispersed aluminium alloy matrix composite pistons has been demonstrated and the laboratory tests have shown promise[9]. The piston being a moving part and the composite pistons being lighter, the envisaged problem in dynamic balancing and non-availability of fatigue property data have slowed down commercialisation. The composite cylinder liners in place of conventional cast iron cylinder liners have shown very promising results in terms of lower damage to piston, lower carbon deposit in the exhaust etc.[9] More field trials are underway. The various graphite particle dispersed aluminium alloy matrix composites shapes/components developed at RRL-T are shown in Fig. 7.

The presence of SiC_p in aluminium alloy matrix showed improvement in the modulus as well as its damping capacity. A typical component being developed and evaluated for vibration testing studies is shown in Fig.8. In addition, extruded tubes and rods of silicon carbide dispersed 2024 and 6061 aluminium alloy matrix composites are also being prepared and evaluated for structural applications.

The non-availability of well recorded published data on AMC in literature and handbooks has inhibited the users to look for AMC for their applications. In addition, standard materials with specified properties are also not available in the market for even trials except very recently from M/s. Duralcan[22], USA, and M/s. Hydro Aluminium, Norway[23]. Even these commercial suppliers do not give any information related to recyclability, and scrap value of the material which in turn affects the cost of the material in general. The composites made by powder metallurgy route are being used in small quantities in the niche areas like aerospace, defence etc. where cost is not the criterion for the selection of the material. However, for general engineering applications stir casting or liquid metal processing route is the cheapest and economically viable[24]. The present trend in the MMC component development is only by trial and error. Even the data on this are not published because the work is mainly carried out by inhouse R&D of the manufacturing companies who have specific interest in commercialising their own products. Under these circumstances the commercialisation of AMC becomes slow and restricts to a few components only. For speedy applications of AMC creation of data banks and standards will certainly help.

CONCLUSION

Regional Research Laboratory, Trivandrum, has been working on the synthesis and evaluation of discontinuous reinforced aluminium alloy matrix composites prepared by stir casting route. The feasibility studies with a few dispersoid systems have identified their limitations and promises. Realising the promises of graphite dispersed composites for self lubricated bodies, short carbon fibre dispersed composites for structural applications and silicon carbide particulate composites for structural and automotive applications, detailed studies were undertaken correlating synthesis, structure and properties. A few components were identified, developed and evaluated in collaboration with the user industries. In the absence of the standards and reliable engineering data on AMC, component development is possible only by prototype evaluations and hence commercialisation becomes very slow. For the speedy applications, creation of data bank and standards are recommended.

ACKNOWLEDGEMENT

The authors wish to acknowledge the support of Dr. A.D. Damodaran Director, Regional Research Laboratory, Trivandrum, and the help of their colleagues of AMC Group during the preparation of this manuscript. The grant-in-aid support received from Aeronautical Research & Development Board (AR&DB) and Department of Science & Technology (DST) New Delhi for the AMC development is also gratefully acknowledged.

REFERENCES

1. P.K. Rohatgi, R. Asthana and A. Das, Int. Met. Review, 1986, Vol. 31, 115-139.

2. R. Mehrabian, R.G. Riek and M.C. Flemings, Metall. Trans., 1974, Vol. 5, 1899-1905.

3. T.P. Murali, M.K. Surappa and P.K. Rohatgi, Metall. Trans., 1982, Vol. 13 B, 485-494.

4. A. Banerjee, M.K. Surappa and P.K. Rohatgi, Metall. Trans., 1983, Vol. 14 B, 273-283.

5. B.N. Keshavaram, K.G. Satyanarayana, B. Majumdar, P.K. Rohatgi and B. Duttaguru, in Proceedings Int. Conf. on Fracture ICF-6, Eds. S.R. Valluri, DMR Taplin, P. Rama Rao, J.F. Knot and R. Dubey, Pergamon Press, New York, 1984, 1979-1987.

6. A. Banerjee and P.K. Rohatgi, J. Mater. Scic., 1982, Vol. 17, 335-342.

7. P.K. Balasubramanian, P. Sreenivasa Rao, B.C. Pai, K.G. Satyanarayana and P.K. Rohatgi, Composite Scic. and Technology, 1990, Vol. 39, 245-259.

8. B.C. Pai, R.M. Pillai, P.K. Biswas, M.C. Shaji and K.G. Satyanarayana, in Proceedings International Conf. on Alumininium INCAL-85, Aluminium Association of India, 1985, Vol. 2, 193-207.

9. B.C. Pai, R.M. Pillai and K.G. Satyanarayana, Indian J. of Engg. & Mater. Scic., 1994, Vol. 1, 279-283.

10. R. Sasikumar and B.C. Pai, in solidification Processing, Institute of Metals, London, 1987, 481-483.

11. B.C. Pai, R.M. Pillai, V.S. Kelukutty, H. Srinivasa Rao, T. Soman, S.G.K. Pillai, K. Sukumaran, K.G. Satyanarayana and K.K. Ravikumar J. Mater. Scic. Letters, 1994, Vol. 13, 1278-1280.

12. B.C. Pai, R.M. Pillai and K.G. Satyanarayana (under preparation)

13. A. Banerjee, S.V. Prasad, M.K. Surappa and P.K. Rohatgi, Wear, 1983, Vol. 82, 141-151.

14. M.C. Shaji, V.S. Kelukutty, B.C. Pai, K.G. Satyanarayana and A.D. Damodaran, in Advances in Composite Materials, Ed. P. Ramakrishnan, Oxford IBH Co. Pvt. Ltd., New Delhi, 1991, 245-252.

15. P.K. Balasubramanian, B.C. Pai, K.G. Satyanarayana and P.K. Rohatgi, in Proceedings advances in Metal Matrix Composites, Eds. M.A. Taha and N.A. El Mahallawy, Trans. Tech. Pub., 1993, 135-144.

16. B.C. Pai, R.M. Pillai, K.G. Satyanarayana and V.S. Kelukutty, Trans. Indian. Inst. Metals, 1992, Vol. 45, 33-43.

17. In Synthesis of Advanced Alloy Matrix Composites, Internal Report, Regional Research Laboratory, Trivandrum, India, 1986.

18. P.K. Rohatgi, S. Ray and Y. Liu, Int. Mater. Review, 1992, Vol. 37, 129-149.

19. H. Sreenivasa Rao, B.C. Pai, P.L. Vinod, R. Manoj and S.S. Sreekumar, Praktical Metallography, 1994, Vol. 31, 190-198.

20. B.C. Pai, K.G. Satyanarayana and P.S. Robi, J. Mater. Scic. Lett., 1992, Vol. 11, 779-781.

21. S. Ciby, B.C. Pai, K.G. Satyanarayana, V.K. Vaidyan and P.K. Rohatgi, J. Mater. Engg. and Performance (ASM), 1993, Vol. 2, 353-358.

22. Duralcan Composites for Gravity Castings - Mechanical and Physical Property Data, 1991, Duralcan composites, San Diego, CA, USA.

23. Aluminium Composites - Mechanical Properties 1992, Hydro Aluminium, Norway.

24. D.J. Lloyd, Inter. Mater. Review, 1994, Vol. 39, 1-23.

25. K.G. Satyanarayana, M.C. Shaji, B.C. Pai, K. Sukumaran, S.G.K. Pillai and A.D. Damodaran, Ind. J. Tech., 1989, Vol. 27, 185-188.

26. B.P. Krishnan, N. Raman, K. Narayanaswamy and P.K. Rohatgi, Wear, Vol. 60 (1980) 205

List of Tables

I   Discontinuous reinforced aluminium alloy matrix composite systems investigated at Regional Research Laboratory, Trivandrum (RRL-T) and their properties.

II   Mechanical properties of the base alloys and 5 vol% graphite dispersed aluminium alloy matrix composites made in the laboratory with different conditions.

III   Tensile properties of the base alloy and short carbon fibre composites hot extruded and fully heat treated (T6) condition in the extruded direction.

IV   Comparison of the tensile properties of the Al-7 Si-Mg-SiC$_p$ composites prepared at RRL(T) and reported in the literature prepared by liquid metal processing route.

Table I

Discontinuous reinforced aluminium alloy matrix composite systems investigated at Regional Research Laboratory, Trivandrum (RRL-T) and their properties

Dispersoid	Size range (μm)	Vol % incorporated	Effect on Tensile properties	Special properties, if any	Components/products	Remarks
1. Coconut shell char	20-250	2-30	Decrease	Adhesive wear resistance above 5 vol%	Bearings	Self lubricated bodies under mild adhesive wear conditions
2. Zircon ($ZrSiO_4$)	5-100	3-30	Improve	Abrasive wear resistance	Pressure die cast self locking water taps, plates	Good wear resistance with about 10 vol% of the particles
3. Zirconia (ZrO_2)	0.01-5	5	–	Abrasive wear resistance	–	–
4. Titania (TiO_2)	0.01-5	5	Improved after hot working	Improved ductility, and high temperature properties	Thin sheets	Easy to form into sheets
5. Flyash	20-150	3-130	Properties improve after hot working	Low density	–	–
6. Glass	20-150	2-200	Properties improve after working	Good corrosion resistance	Rods	Can be used for structural applications
7. Graphite particles	5-250	2-50	Decrease	Adhesive wear resistance	Contactors, bearing pistons, cylinder liners etc.	Self lubricated bearings
8. Short carbon fibre	7 μm φ	5-15	Improve after extrusion	High modulus & Strength	Rods, tubes	Structural applications
9. Silicon carbide particles	15-45	5-35	Improve	Abrasive wear resistance, high temp. properties	Tubes, vibrator components	"

Table - II.

Mechanical properties of the base alloys and 5 vol.% graphite dispersed composites cast under different conditions

Material	Condition	Temper	UTS (MPa)	Elongation (%)	Hardness BHN
Al-12Si-Mg (LM6)	Gravity Die cast	As cast	160	2-3	70-80
Al-12Si-Mg+ Graphite	"	"	145	2	70-80
Al-7Si-Mg	"	"	135	2-3	70
Al-7Si-Mg	"	T6	300	6	100
Al-7Si-Mg+ Graphite	"	As cast	100	2	70-80
Al-7Si-Mg + graphite	"	T6	235	2-3	105
Al-7Si-Mg	pressure Die cast	As cast	145	4	75
Al-7Si-Mg + graphite	"	"	135	2	85-95
Al-11Si-Mg-Fe (LM13)	"	"	165	2	85-90
"	"	T6	295	3-4	-
Al-11Si-Mg-Fe + graphite	"	As cast	135	2-3	-
Al-11Si-Mg-Fe + graphite	"	T6	265	2-3	-
Al-16.5 Si-Mg	"	As cast	240	1-2	110-120
Al-16.5 Si-Mg + + graphite	"	"	225	1-2	115-125

Table - III.

Tensile properties of base alloy and short carbon fibre composites, hot extruded and fully heat treated (T6) condition, in the extruded direction.

Material	UTS (MPa)	0.2% Proof strength (MPa)	Elongation (%)	Modulus (MPa)
Base alloy (Al-6.2 Zn-2.45 Mg- 1.7 Cu-0.15 Zr.)	480-490	445-455	10-12	70
Base alloy + 5 wt% Fibre	425-480	435-460	2-4	75-80
Base alloy + 8 wt% Fibre	500-510	490-500	2-3	80-85

Table - IV.

Tensile properties of the composites prepared at
RRL-T and those reported in the literature prepared by
liquid metal processing route

Organisation and Base alloy	Vol% SiC_p	SiC_p size (μm)	UTS (MPa)	Elongation (%)	Modulus GPa
RRL-T Al-7Si-Mg	10	42	280-310	2.0	81
	15	"	280-320	2.0	90
	20	"	305-365	1.0	105
Hydro Aluminium Corporation Al 7 Si Mg	15	9-23	308	0.6	91
M/s Duralcan USA	10	10-12	308	0.6	82
Al 7 Si Mg	15	"	336	0.3	91
	20	"	357	0.4	98

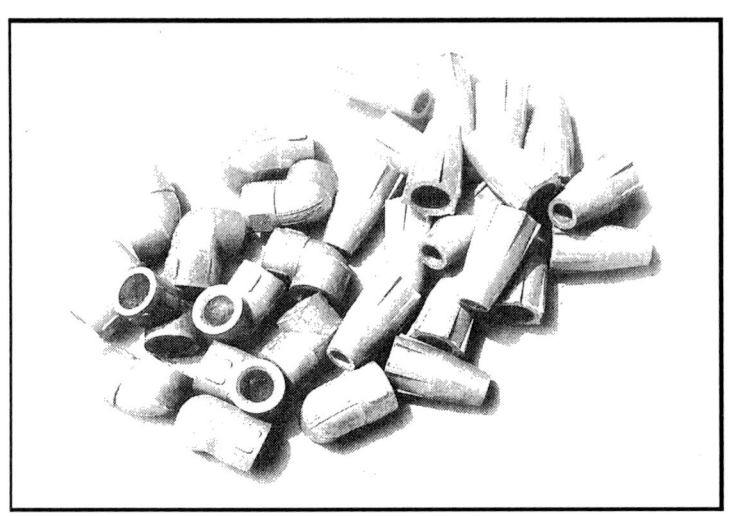

Fig. 1: Self locking water taps made by pressure die casting with aluminium-15 wt% zircon composite

Fig. 2: Typical master composite ingot containing about 50 vol% graphite made by semi solid slurry casting technique

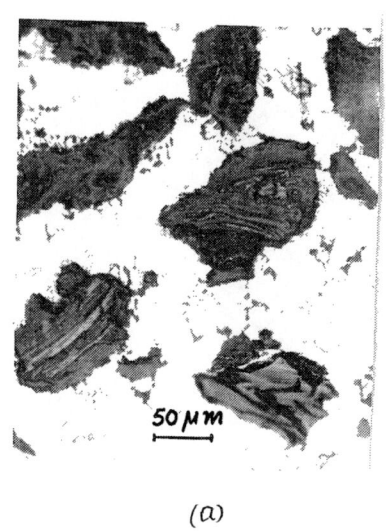

(a)

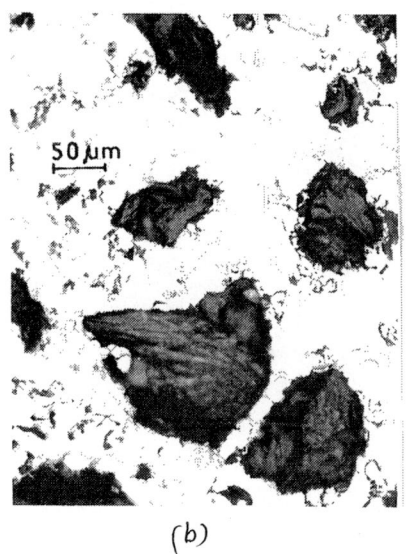

(b)

Fig. 3: Typical microstructure of 5 vol% graphite dispersed
Al-7 Si-Mg alloy matrix composite prepared by
   (a) Dilution of the master alloy
   (b) Direct liquid metal processing technique

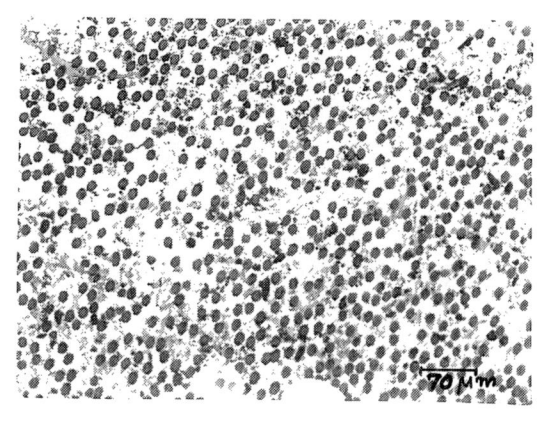

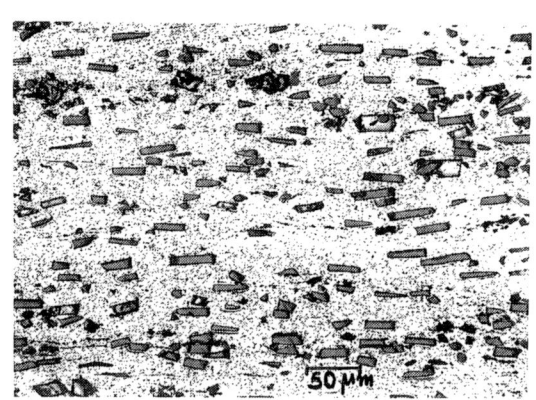

Fig. 4: Microstructures of 8 wt% short carbon fibre dispersed hot extruded composite

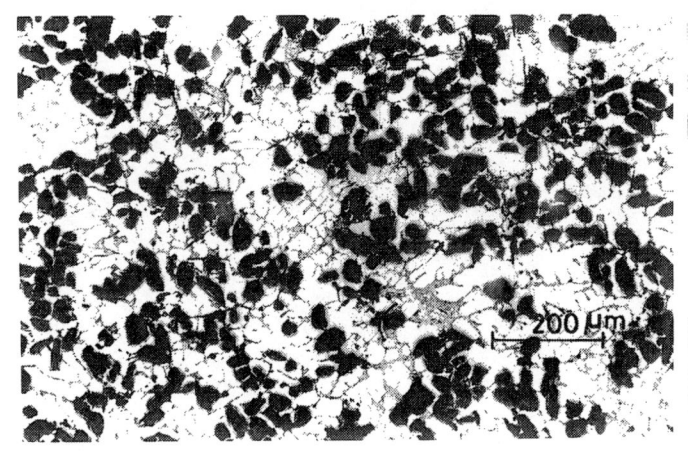

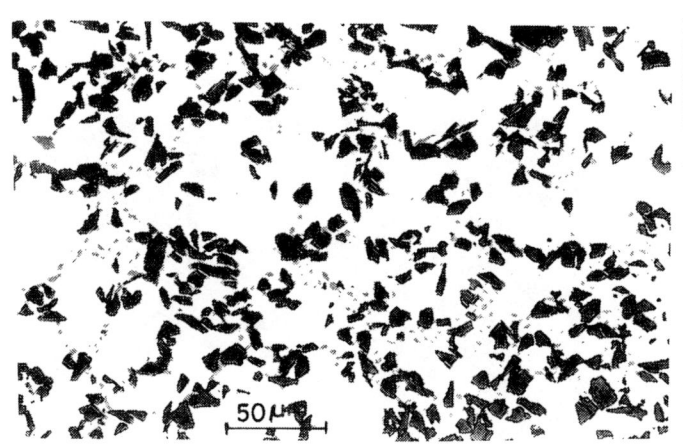

Fig. 5: Comparison of the 20 vol% SiC_p dispersed aluminium alloy matrix composite
(a) Prepared at RRL-T with 43 um APS, SiC_p
(b) Given in M/s. Duralcan literature[22]

Fig. 6: Contactors made out of Al-5 vol% graphite composites by pressure die casting

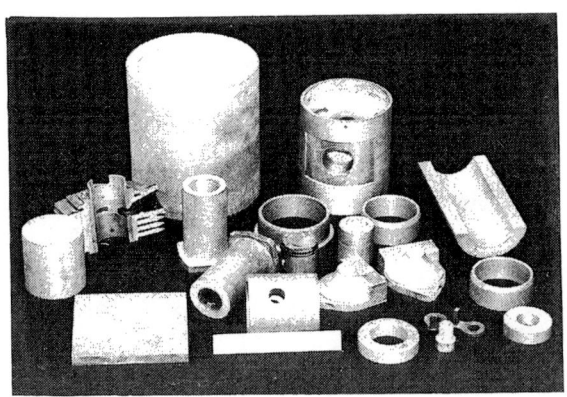

Fig. 7: The various components made with graphite dispersed Al alloy matrix composites

Fig. 8: A component made for vibration application with Al-12 Si-Mg alloy SiC_p composite by sand casting

Structure-Property Relationships for Layered DRA Materials

Todd M. Osman[1], John J. Lewandowski[1], Warren H. Hunt, Jr.[2], and Donald R. Lesuer[3]

[1] - Department of Materials Science and Engineering, The Case School of Engineering, Case Western Reserve University, Cleveland, OH
[2] - Alcoa Technical Center, Alcoa Center, PA
[3] - Lawrence Livermore National Laboratory, Livermore, CA 94551

Discontinuously reinforced aluminum (DRA) materials are being developed for structural applications based upon their elevated specific stiffness and strength properties. The utilization of these materials for such applications, however, is often hindered by the degradation in fracture resistance for DRA materials when compared to monolithic metallic alloys. Much of the literature on fracture toughness improvements in DRA materials has focused on intrinsic modifications of the microstructure (*e.g.*, particle size, reinforcement type, and matrix condition).[1-7] While enhancements in fracture toughness may be achieved via such a route, a potentially more fruitful method for damage tolerance improvements may be through the use of extrinsic toughening mechanisms.

The goal of extrinsic toughening techniques is to reduce the driving force for crack propagation by providing alternate, non-catastrophic crack propagation routes or via a reduction in the local stress intensity at the crack tip.[8] These techniques have been extensively utilized to produce toughness improvements in brittle materials (*i.e.*, ceramics and intermetallics)[9-15] through the use of brittle/ductile laminates or ductile phase toughened materials. Recently, similar techniques have been employed to increase the fracture resistance of DRA materials.[3,6,16-29] In this case, regions of monolithic aluminum are combined with a DRA material to produce a hybrid structure. While such "ductile phase toughening" of a DRA material may be similar to that in brittle/ductile systems[9-15], the two cases are not completely analogous due to the inherently greater fracture resistance of a DRA material when compared to that of a ceramic or intermetallic (*e.g.*, 18-30 MPa$\sqrt{m}$ for representative DRA materials[6] versus 3-8 MPa$\sqrt{m}$ for representative ceramics and intermetallics[30]) as well as lower property mismatches between the constituent materials[27].

Recent work[16-29] has shown that the fracture resistance of a DRA material may be greatly improved via the use of such extrinsic toughening techniques. The purpose of this paper is to review the levels of fracture toughness improvements achievable in layered DRA materials. In particular, recent efforts in the development of DRA laminates are presented. Important trends in the fracture resistance are discussed in light of the relevant structure-property relationships for layered structures.

DRA Laminates

DRA laminates consisting of alternating layers of monolithic aluminum and a DRA material (see Figure 1) are being developed to utilize extrinsic toughening mechanisms on a macroscopic level for toughness improvement. A potential application for these materials is as aerospace plate product. In such applications, two loading directions in Figure 1 are important. In the crack arrestor orientation, crack growth occurs normal to the interfaces. In the crack divider orientation, crack propagation occurs parallel to the bondplane; thus, crack growth may proceed simultaneously in all layers. In a plate product in which the thickness is much less than the width and length, damage in the crack arrestor orientation should occur due to impact loadings. If a static applied Mode I loading is assumed, damage in a DRA laminate is most likely from an edge or rivet hole in the crack divider orientation. Based upon these loading conditions, the impact resistance of a DRA laminate in the crack arrestor orientation and the fracture resistance in the crack divider orientation must be considered.

Table I compares the impact resistance for monolithic aluminum alloys, DRA materials, and DRA laminates in the crack arrestor orientation. As shown previously[29], the impact resistances of the DRA materials are much less than those of the monolithic aluminum alloys. The impact resistance of a DRA laminate not only exceeds that of the constituent DRA material, but can be greater than a monolithic aluminum alloy. This large improvement in fracture resistance is directly related to the presence of a continuous interface between the DRA material and the monolithic aluminum alloy in the DRA laminate.

In order to analyze crack growth during impact loading, three-point bend specimens have been tested to determine the crack growth mechanisms.[6,22] Fracture of a typical DRA material is concentrated in a localized region and a planar fracture surface is produced. By contrast, extensive non-planar crack growth and crack blunting can occur in the crack arrestor orientation via interfacial delamination (see Figure 2). After crack arrest, continued propagation can only occur after re-initiation on the tensile surface of subsequent layers in a manner analogous to an unnotched bend bar. The degree of toughening in such systems, therefore, is directly proportional to the length of delamination.[16,17,22,28] The lower fracture toughness of the adhesively bonded 7093/SiC/15p-7093 laminate in Table 1 when compared to the roll bonded 7093/SiC/15p-7093 laminate may be related to the lack of interfacial delamination observed in that material.[22] In this case, improvements in impact resistance occur solely due to energy absorption via plastic dissipation in the monolithic aluminum alloy.

In addition to improvements in impact resistance, DRA laminates have been found to exhibit enhanced Mode I fracture resistance. Figure 3 illustrates the reduction in fracture toughness for DRA materials as the reinforcement content increases. For a monolithic DRA material, catastrophic fracture often occurs at the maximum load without any stable cracking.[4] In this case, the initiation toughness[31], K_q, is a sufficient measure of fracture resistance. By contrast, improved fracture resistance is produced in DRA laminates due to the establishment of R-curve behavior in which stable, non-catastrophic cracking occurs after initiation.[3,18-23,26,27] Figure 4 shows a schematic of the crack growth profile in DRA laminates in which damage extends further in the DRA layers than in the monolithic aluminum layers. The preferential cracking of the DRA material results in a substantial uncracked aluminum ligament in the crack wake. In a recent study[23], it was found that stabilization of cracking in the DRA layer is strongly related to the plastic dissipation of energy in the aluminum ligaments via a crack bridging mechanism.

Due to the large extent of non-planar crack growth in the crack divider orientation, it is difficult to utilize a J-integral approach[32] to quantify crack growth resistance for the DRA laminates. In order to quantify the total fracture resistance of the laminate structures in Figure 3, an apparent fracture toughness, K_{ee}, was utilized based upon the equivalent energy calculation in ASTM E992[33]. The laminates in Figure 3 represent an improvement in fracture resistance over a monolithic DRA material on the basis of effective global silicon carbide volume fraction. If one considers the fact that the stiffness of a DRA material is predominantly a function of the volume fraction of reinforcement[34], Figure 3 suggests that a DRA laminate approach affords the potential for enhanced stiffness-toughness combinations in DRA materials.

As can be seen in Figure 3, the fracture toughness of a DRA laminate is not solely a function of effective global silicon carbide volume percent. Just as in the above discussion about fracture in crack arrestor orientation, interfacial delamination will influence fracture resistance in the crack divider orientation. For the 7093/SiC/15p-7093 OA system[22], the fracture resistance was found to be directly proportional to the extent of interfacial delamination (*i.e.*, greatest with the largest amount of interfacial delamination). By contrast, in the MB78/SiC/20p-7093-OA laminates[27], the persistence of interfacial delamination served to reduce the growth toughness. In the case of 7093/SiC/15p-7093-OA laminates, the 7093 layer was relatively thick (2.5 mm) when compared to the 7093 layer thickness in the MB78/SiC/20p-7093 OA laminates (1.5, 0.90, and 0.45 mm). The apparently contradictory trends in growth toughness with changes in interfacial delamination in the two DRA laminate systems may be understood by a consideration of the influence of sample thickness on toughness for aluminum sheet material.

Figure 5 displays such a relationship for another high strength aluminum alloy, 7075. The influence of thickness on the fracture toughness for 7093 is expected to be relatively similar to that shown for 7075 in Figure 5. The 2.5 mm 7093 in the 7093/SiC/15p-7093-OA laminates should therefore be near the peak toughness for 7093. In this case, interfacial delamination will be beneficial, with the laminate benefitting from the enhanced toughness of the aluminum. By contrast, the 1.5 mm, 0.90 mm, and 0.45 mm layer thicknesses utilized in the MB78/SiC/20p-7093-OA laminates would be in the regime where the toughness of 7093 is expected to decrease with decreasing layer thickness due to net section yielding. Delamination, therefore, may be detrimental in this regime as it relieves the constraint on the aluminum and reduces the load bearing capacity of the aluminum layer.[27] This suggests that if the thickness of the monolithic aluminum is to the right of the peak toughness on a toughness versus thickness plot (see Figure 5), interfacial delamination is favorable due to the elevated toughness of the monolithic aluminum; however, if the aluminum layer thickness is to the left of the peak toughness on such a plot, interfacial delamination may be detrimental due to the reduced load bearing capacity of the aluminum ligament.

Summary

The toughness of DRA materials may be markedly improved via the use of extrinsic toughening mechanisms. DRA laminates consisting of alternating layers of monolithic aluminum and DRA materials have been found to provide an increased fracture resistance via the production of stable, non-catastrophic crack growth. In the crack arrestor orientation, impact resistance improvements are directly related to production of crack blunting and arrest via interfacial delamination and plastic dissipation in the monolithic aluminum layers. In the crack divider orientation, two dissimilar relationships between interfacial delamination and Mode I fracture toughness may exist. If one considers the toughness-thickness relationship for monolithic aluminum, interfacial delamination may be beneficial in the plane strain to plane stress transition regime due to the increase in toughness with a decrease in thickness. By contrast, at thin layer thicknesses, interfacial delamination may be detrimental due to the decrease in load bearing capacity of aluminum with decreasing thickness in this regime.

Acknowledgments

The authors would like to acknowledge the collaboration of Dr. Chol Syn and Dr. Robert Riddle of Lawrence Livermore National Laboratory and Dr. Ralph Bush and Dr. Jan Teply of Alcoa Technical Center. Finally, the support of Alcoa, NSF-DMR-PYI-89-58326 (JJL,TMO), and U.S. Department of Energy (LLNL) through contract No. W-7405-Eng-48 (DRL,TMO) is acknowledged.

References

1. F.J. Humphreys, *Mechanical and Physical Behavior of Metallic and Ceramic Composites,* ed. S.I. Anderson, H. Lilholt, and O.B. Pederson, Riso National Laboratory, Rosklide, Denmark, September 05-09, 1988, p. 51.

2. Y. Flom and R.J. Arsenault: *Acta metall.*, 1989, vol. 37, p. 2413.

3. J.J. Lewandowski, C. Liu, and W.H. Hunt, Jr., *Mat. Sci. and Eng.*, v. A107, 1989, p. 241.

4. M. Manoharan and J.J. Lewandowski, *Acta metall.*, V. 38, 1990, p. 489.

5. D.L. Davidson, Metal Matrix Composites: Mechanisms and Properties, ed. R.K. Everett and R.J. Arsenault, Academic Press, Boston, 1991, p. 217.

6. W.H. Hunt, Jr., T.M. Osman, and J.J. Lewandowski, *JOM*, v. 45, January 1993, p. 30.

7. D.J. Lloyd, *Acta metall.*, v. 39, 1991, p. 59.

8. R.O. Ritchie, *Mat. Sci. Eng.*, v. A103, 1988, p. 15.

9. V.V. Kristic, P.S. Nicholson, and R.G. Hoagland, *Journal of the American Ceramic Society*, v. 64, 1981, p. 499.

10. M. Mendiratta, J.J. Lewandowski, and D.M. Dimiduk, *Metall. Trans. A.*, v. 22A, 1991, p. 1573.

11. P. Mataga, *Acta metall.*, v. 37, 1989, p. 3349.

12. J. Kajuch, J.D. Rigney, and J.J. Lewandowski, *Mat. Sci. Eng.*, v. A115, 1992, p. 59.

13. H.C. Cao and A.G. Evans, *Acta metall.*, v. 39, 1991, p. 2997.

14. M. Bannister and M.F. Ashby, *Acta metall.*, v. 39, 1991, p. 2373.

15. J. Kajuch, J. Short, and J.J. Lewandowski, *Acta metall.*, v. 43, 1995, p. 1955.

16. V.C. Nardone, J.R. Strife, and K.M. Prewo, *Metall.Trans. A*, v. 22A, 1991, p. 171.

17. F. Zok, S. Jansson, A.G. Evans, and V. Nardone, *Metall.Trans. A*, v. 22A, 1991.

18. T.M. Osman, J.J. Lewandowski, D.R. Lesuer, C.K. Syn, and W.H. Hunt, Jr., Aluminum Alloys: Their Physical and Mechanical Properties (ICAA4), v. II, eds. T.H. Sanders, Jr. and E.A. Starke, Jr., The Georgia Institute of Technology, Atlanta, GA, 1994, p. 706.

19. M. Manoharan, L. Ellis, and J.J. Lewandowski, *Scripta Metall.*, v. 24, 1990, p. 1515.

20. L. Yost Ellis and J.J.Lewandowski, *J. Mat. Sci. Letters*, v.10, 1991, p. 461.

21. L. Yost Ellis and J.J. Lewandowski, *Mat. Sci. Eng.*, v. A183, 1994, p. 59.

22. T.M. Osman, J.J. Lewandowki, and W.H. Hunt, Jr., *Metall. Trans. A*, 1995, in review.

23. T.M. Osman, P.M. Singh, and J.J. Lewandowski,*Scripta Metall.*, v. 10, 1994, p. 607.

24. C.K. Syn, D.R. Lesuer, K.L. Caldwell, O.D. Sherby, and K.R. Brown, <u>Developments in Ceramic and Metal-Matrix Composites</u>, ed. K. Upadhya, TMS, Warrendale, PA, 1992, p. 311.

25. C.K. Syn, D.R. Lesuer, and O.D. Sherby, <u>Light Materials for Transportation</u>, ed. N.J. Kim, Kyongju, Korea, 1993, p. 763.

26. D. Lesuer, C. Syn, R. Riddle, and O. Sherby, <u>Intrinsic and Extrinsic Fracture Mechanisms in Discontinuously Reinforced Composites</u>, ed. J.J. Lewandowski and W.H. Hunt, Jr., TMS, Warrendale, PA, 1995, p. 93.

27. T.M. Osman, J.J. Lewandowski, W.H. Hunt, Jr., D.R. Lesuer, and R. Riddle, <u>Intrinsic and Extrinsic Fracture Mechanisms in Discontinuously Reinforced Composites</u>, ed. J.J. Lewandowski and W.H. Hunt, Jr., TMS, Warrendale, PA, 1995, p. 103.

28. V.C. Nardone, <u>Intrinsic and Extrinsic Fracture Mechanisms in Inorganic Composite Systems,</u>, eds. J.J. Lewandowski and W.H. Hunt, Jr., TMS, Warrendale, PA, 1995, p. 85.

29. J.J. Lewandowski and P.M. Singh, <u>Intrinsic and Extrinsic Fracture Mechanisms in Discontinuously Reinforced Composites</u>, ed. J.J. Lewandowski and W.H. Hunt, Jr., TMS, Warrendale, PA, 1995, p. 129.

30. A.G. Evans and K.T. Faber, <u>Fracture in Ceramic Materials</u>, ed. A.G. Evans, Noyes Publications, Park Ridge, NJ, 1984, p. 109.

31. ASTM E399, ASTM, Philadelphia, PA, 1983.

32. ASTM E813, ASTM, Philadelphia, PA, 1992.

33. ASTM E992, ASTM, Philadelphia, PA, 1989.

34. T.M. Osman, J.J. Lewandowski, and W.H. Hunt, Jr., <u>Advances in Powder Metallurgy and Particulate Materials-1994</u>, eds. C. Lall and A. Neupaver, Metal Powder Industries Federation, Princeton, NJ, 1994, p. 351.

35. T.M. Osman and J.J. Lewandowski,*Scripta Metall.*, v. 31, 1994, p. 191.

36. D.S. Thompson, *Metallurgical Transactions A*, v. 6A, 1975, p. 671.

Table I: Impact resistance for monolithic aluminum alloys, DRA materials, and crack arrestor DRA laminates.

Material	DRA Thickness (mm)	Aluminum thickness (mm)	Energy absorbed (J/cm^2)
MB85 - UA	-	10	28.3
MB85/SiC/15p - UA	10	-	3.2
Press bonded laminate	7.2 (one layer)	2.8 (one layer)	40.4
Press bonded laminate	6.0 (one layer)	4.0 (one layer)	56.0
7093-T7E92	-	10	8.2
7093/SiC/15p-T7E92	10	-	2.0
Roll bonded laminate	5.0 (one layer)	2.5 (two layers)	28.4
Adhesively bonded laminate	5.0 (one layer)	2.5 (two layers)	9.5

Alloy Chemistry:		
MB85:	Al-3.5 wt/o Cu-1.5 wt/o Mg-0.4 wt/o Zr-0.21 wt/o Mn	
7093:	Al-9.0 wt/o Zn-2.2 wt/o Mg-1.5 wt/o Cu-0.14 wt/o Zr-0.10 wt/o Ni	
MB85/SiC/15p:	MB85 with 15 volume percent silicon carbide particles	
7093/SiC/15p:	7093 with 15 volume percent silicon carbide particles	

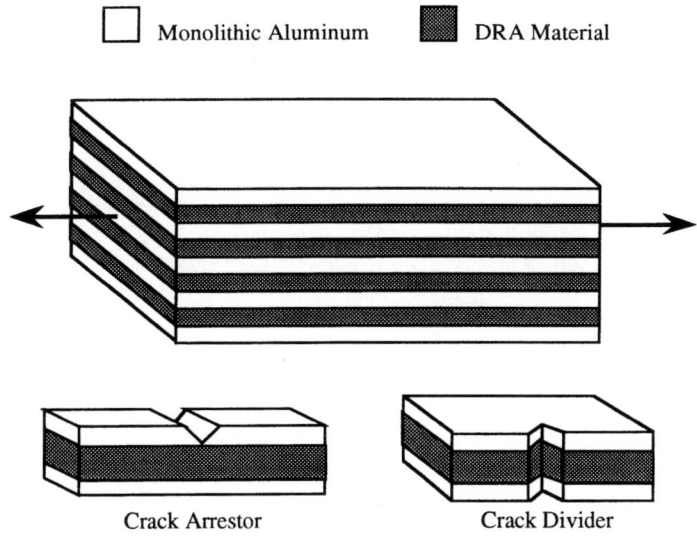

Figure 1:   DRA Laminates

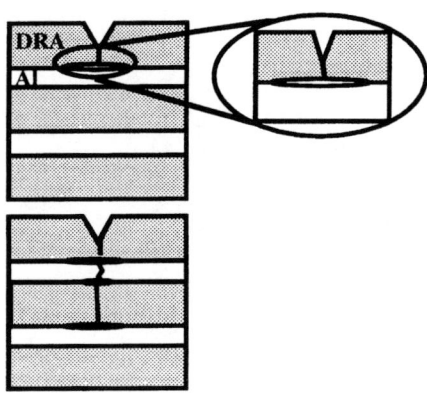

Figure 2:   Crack growth in the crack arrestor orientation of a DRA laminate.[18,22]

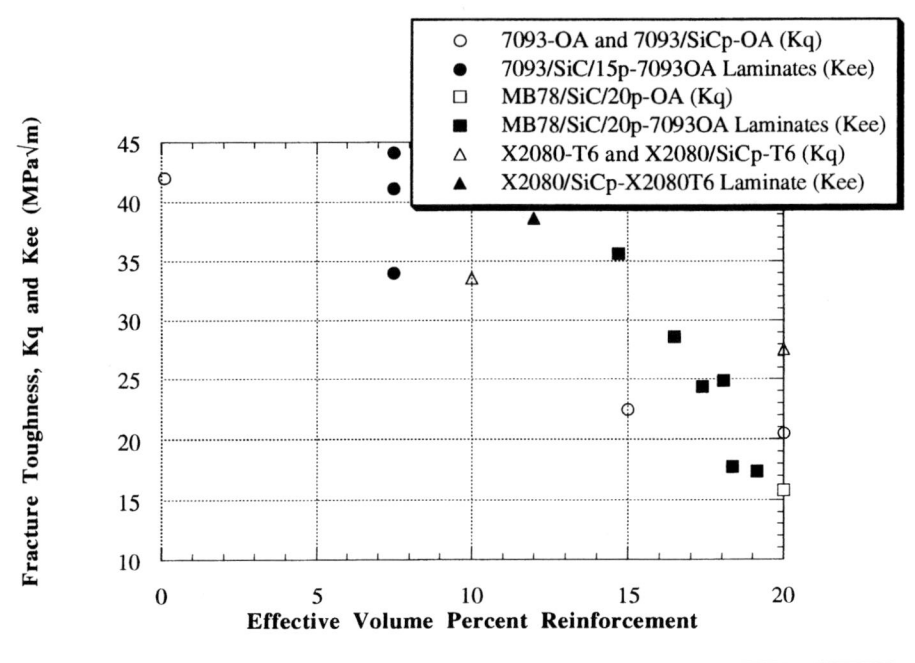

Alloy Chemistry: 7093: Al-9.0 wt/o Zn-2.2 wt/o Mg-1.5 wt/o Cu-0.14 wt/o Zr-0.10 wt/o Ni
MB78: 7.0 wt/o Zn, 2.0 wt/o Cu, 0.14 wt/o Zr
X2080: Al-3.8 wt/o Cu-1.8 wt/o Mg-0.2 wt/o Zr

7093/SiCp: 7093 with silicon carbide particles
MB78/SiCp: MB85 with silicon carbide particles
X2080/SiCp: X2080 with silicon carbide particles

7093/SiC/15p-7093 OA: Three layers (Roll bonding and Adhesive bonding)[22]
MB78/SiC/20p-7093 OA: Three layers (Press bonding)[27]
X2080/SiC/20p-X2080 T6: Five layers (Adhesive bonding)[34,35]

Kq : Calculated based upon ASTM E399.[31]
Kee: Calculated based upon ASTM E992.[33]

Figure 3: Fracture toughness of aluminum, DRA materials, and DRA laminates as a function of effective volume percent of silicon carbide particles.

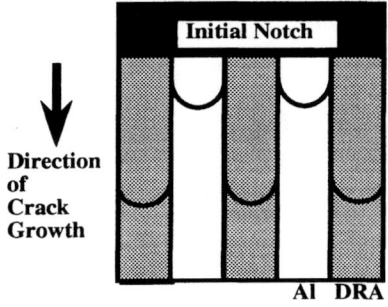

Figure 4: Crack growth in the crack divider orientation.[6,18,22,26,27,34]

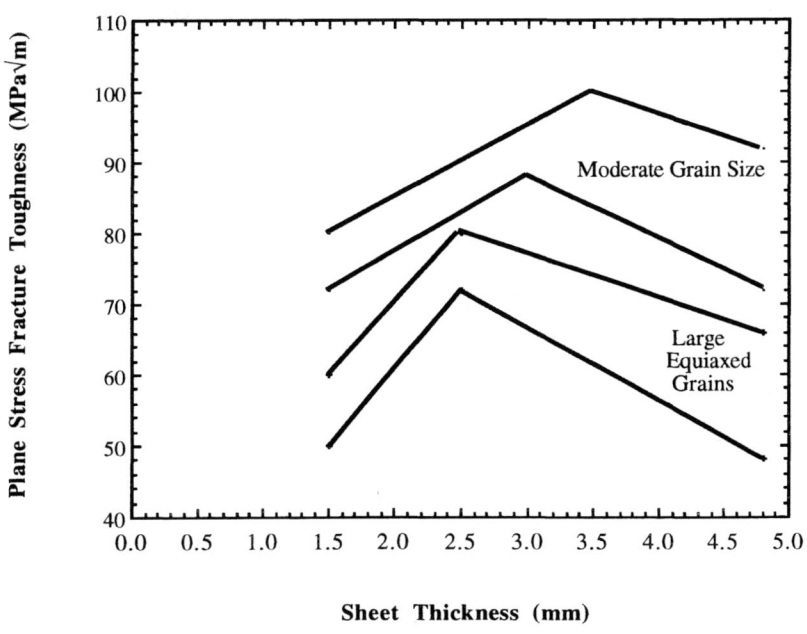

Figure 5: Dependence of toughness on thickness for 7075 (Al-5.6 wt/o Zn-2.5 wt/o Mg-1.6 wt/o Cu-0.23 wt/o Cr).[36]

Processing of a Particulate Metal Matrix Composites by Roll Bonding Method

Y.P. YAO and W.B. LEE

The manufacture of an aluminium alumina composite sheet from anodising of aluminium foils and roll bonding method are described. The process can be automated for continuous production of PMMC in sheet form. The microstructures of the materials are examined. Cracks introduced in the materials from the breaking up of oxide is one of the main cause of porosity in the rolled materials. The porosity in the hot rolled material can be reduced by further cold rolling.

I. INTRODUCTION

There have been extensive researches in the manufacture of Particulate Metal Matrix Composites (PMMC) over the pass forty years. Two of the most commonly used techniques to produce PMMC are powder metallurgy and casting [1,2]. In this paper, a method of producing Al-Al_2O_3 particulate MMC materials by hot rolling [3] is used. The main techniques include anodising of aluminium foils and hot roll bonding which can be automated for continuous production of PMMC in sheet form. These materials have good electrical conductivity and creep properties. To optimise the processing route, an investigation is carried out to study the effect of some of the processing variables on the microstructures and mechanical properties of the PMMC.

II. EXPERIMENTAL PROCEDURES

A schematic diagram of the processes involved in the manufacture of Al-Al_2O_3 composites is shown in Fig. 1. The major processes involved are electrolytic anodising and hot-roll bonding process.

Firstly, commercially pure (CP) aluminium foils were degreased in a detergent bath followed by an acidic etching with nitric acid (50%) to remove the grease or oil left on the aluminium foils from previous handling and fabricating processes. These treatments preceded immediately before the anodising process.

The aluminium foils were anodised in a 20 vol% of sulphuric acid to produce an aluminium oxide film on the surface. During the anodising process, the foils were placed directly on the electrode without any racking fixture to eliminate the accumulation of anodic film on the contact area between the rack and the foil. The size of each foil was kept the same in the anodising bath. The rate of the oxide film formation was found to be 0.5 μm /min and an oxide thickness of 5 μm was obtained under the above condition. The Al_2O_3 film thus obtained was hydrated due to its porous structure. Sealing process was carried out to close the pores. The corrosion and wear resistance of the oxide film are improved by sealing. The main steps in anodising are listed in Table I.

Hundreds of the anodised foils were stacked and sandwiched with plain aluminium foils. Different volume fractions of alumina were achieved by altering the proportion of anodised foils to plain foils. The stack of the foils was consolidated for an hour under a hydraulic pressure of 150 bar and at a temperature of 150°C. Hot rolling was then carried out at 550°C to break up the oxide film into particulates and bonding between the matrix and the alumina were facilitated by mechanical keying.

TABLE I. The main steps in anodising

Steps:	Solvent:	Duration:	Temperature:	Remarks:
I) Chemical treatment				
a) Cleaning (degreasing)	Detergent	5min	80°C	Remove the grease and oil
b) Desmudging	50%vol HNO_3	20s	Room temp.	Remove the inorganic dirt
II) Anodising	20%vol H_2SO_4	10min	Room temp.	Current Density:1.5A/dm2 Anode to cathode size ratio : 1:1
III) Post-treatment				
a) Sealing	Distilled water	10min	100°C	Result in anhydrous alumina.
b) Oven dry	N/A	10min	100°C	Dry the foils

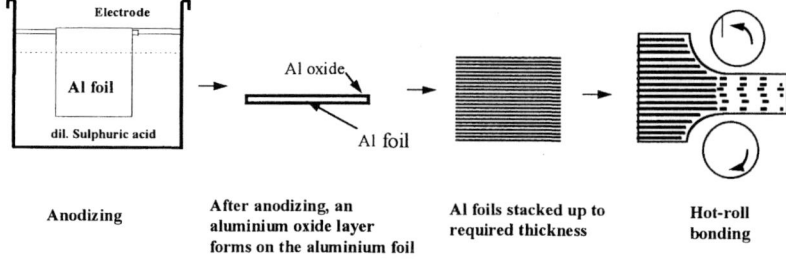

Figure 1. Schematic diagram showing the manufacturing routes of the Al-Al$_2$O$_3$ MMC.

The specimens were annealed at 500°C for an hour and cold rolled to reinforce the bond. All the specimens were rolled in the same direction. The composites were annealed again at 500°C for an hour.

Table II. Experimental parameters for material preparation.

Specimen No.	Volume fraction	Roll Bonding			
		Hot roll (%)	Int. anneal temp. (°C)	Cold roll (%)	Final anneal temp. (°C)
1	~5%	90	N/A	0	N/A
2		90	500	0	N/A
3		90	500	30	N/A
4		90	500	30	500
5		90	500	60	N/A
6		90	500	60	500
7	~0%	65	500	80	500
8	~5%	65	500	80	500
9	~15%	65	500	80	500

PMMC sheets were prepared according to Table II. The microscopic and mechanical properties of the PMMC materials with various cold-roll reductions were evaluated, both in the as-rolled and annealed conditions (Specimen No. 1-6). Three different volume fractions of alumina content were used to assess the properties affected by the addition of reinforcement content (Specimen No. 7-9). An unreinforced matrix material, i.e., plain aluminium foils (Specimen 7) were prepared from the same bonding method as for the anodising foils.

The volume fractions of alumina were examined by metallograph inspection. The alumina volume fraction V$_r$ in the PMMC sheets from both the longitudinal section and the transverse section was measured with an image analysis software (Quantimet 500).

The microstructures of the MMCs were examined with optical microscope. The morphology of the fracture surface was inspected using a scanning electron microscope. A Vickers microhardness tester was used to measure the hardness of the composites with a load of 100g. In order to eliminate possible segregation effects, a minimum of ten random hardness readings was taken for each specimen.

III. RESULTS AND DISCUSSIONS

The optical micrographs at both the longitudinal and transverse sections of the specimen were shown in Fig. 2. The distribution of the particulates is homogeneous at the optical level. No significant particle-free bands are observed. The alumina has a rectangular platelet shape with an average thickness of 5µm. The aspect ratios (ratio of length to thickness from both sectional planes) are found to be higher in the transverse section than those in the longitudinal section. The compression pressure from the rolling process was sufficient to break up the hard oxide layer into a uniform dispersion of fine rectangular platelets. The oxide tends to breaks up easier in a direction perpendicular to the rolling direction in accordance with the observation made by Lee and Subramanian[4] in aluminium alumina particulate composites prepared by casting. The tensile properties of the MMC materials have also been studied. The tensile strength and the fracture surface (Fig. 3) were affected by the shape and the orientation of the platelets.

The volume fractions of the MMC materials have been measured and the results are shown in Table III. The measured values from quantitative metallograpy are close to the theoretical ones.

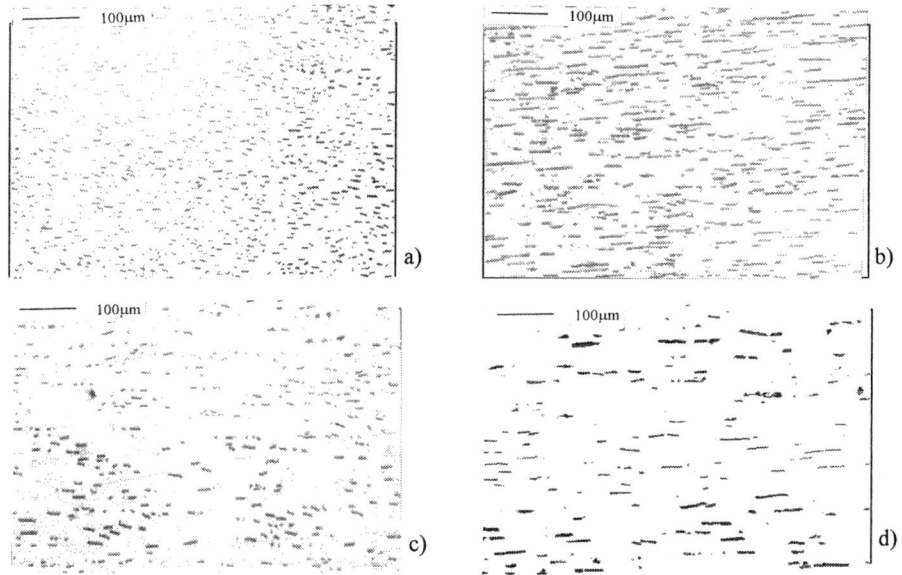

a)15%, longitudinal section; b)15%, transerve section; c)5%, longitudinal section; d)5%, transverse section
Figure 2. Volume fraction of Al_2O_3 MMC.

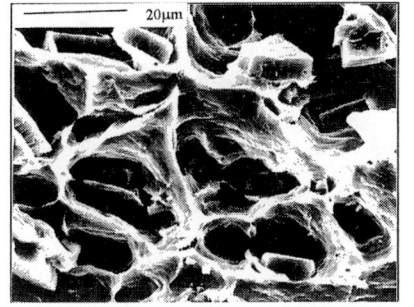

a)   Fracture surface in longitudinal section.   b)   Fracture surface in transverse section
Figure 3. Fracture surfaces showing different fracture observed from different sections.

TABLE III. The volume fraction of the MMC materials

Specimens	1	2	3	4	5	6	7	8	9
V_r (theoretical), %	5	5	5	5	5	5	0	5	15
V_r (quantitative metallography), %	5.84	6.23	6.19	5.27	6.10	6.61	0	6.09	14.5
V_r (density measurement), %	1.73	<0	0.06	<0	0.54	<0	0	0.34	1.74

The volume fractions of the alumina content (V_r), from the density measurement are calculated by,

$$V_r = \frac{\rho_c - \rho_m}{\rho_f - \rho_m} \quad \ldots\ldots\ldots(1)$$

where ρ_c, ρ_m and ρ_f are the density of the PMMC materials, the aluminium matrix and the density of the alumina respectively. ρ_m and ρ_f are taken to be 2.68 gcm^{-3} and ρ_f =3.9 gcm^{-3}. A disagreement is found between the volume fraction determined by quantitative metallography and by density measurements. Such a difference can be attributed to the porosity which are present in the material. In this case, V_r should be corrected as,

$$V_r = \frac{\rho_c - \rho_m(1 - V_p)}{\rho_f - \rho_m} \quad \ldots\ldots(2)$$

where V_p is the porosity fraction.

Assuming that the volume fraction of Al$_2$O$_3$ present in the composites is the theoretical volume fraction, values for V_p are found to be related to the cold rolling reduction (Fig. 4). The porosity fraction is found to be reduced as the percentage of cold rolling reduction increases. Cold rolling can improve the interface bonding between the matrix and the alumina.

Two types of micro-cracks are found in the composites after cold rolling. A schematic diagram of the crack is shown in Fig. 5. The first type of crack occurs at the interface between the alumina and aluminium matrix due to the breaking up of alumina platelets with unfilled gaps in the matrix (Fig. 5a). These pore-like cracks often appear at the two ends of the rectangular platelets. The second common defect is the penny-shaped cracks appeared in the mid-plane of the rectangular platelets as shown in Fig. 5b. These cracks will reduce the overall mechanical properties [5].

The results of the Vickers hardness tests are shown in Fig. 6. The hardness is found to increases linearly with the amount of cold rolling reduction. The effect of the annealing process on the hardness variation is also shown in Fig.6.

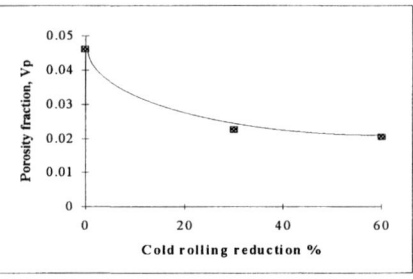

Figure 4. The effect of cold rolling reduction on the calculated porosity fraction

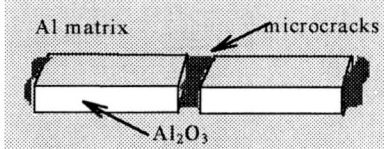

a) Interfacial debonding
(pore-like cracks)

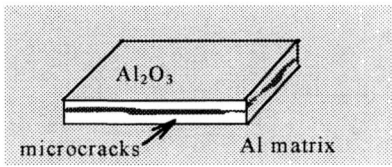

b) Penny-shaped crack

Figure 5. Two types of internal defects induced from the rolling process.

IV. CONCLUSION

A simple method to produce Al-Al$_2$O$_3$ PMMC material by a hot-rolled bonding is described. The process can be highly automated to produce PMMC in sheet form. The volume fraction of the Al$_2$O$_3$ can be controlled by varying the oxide thickness and the stacking sequence of the anodised and plain foils before hot rolling. By applying a further cold roll reduction to the hot rolled PMMC sheet, the volume fraction of porosity can be reduced.

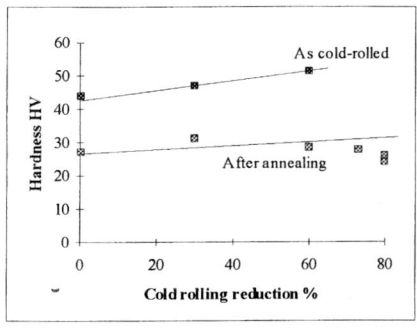

Figure 6. Relation between cold rolling reduction and Vickers hardness. The hardness of the annealed specimen is also shown.

V. ACKNOWLEDGEMENT

The authors would like to thank the Research Committee of the Hong Kong Polytechnic University for the award of a research studentship to Y.P.Yao to conduct this research.

VI. REFERENCES

1. M.G.Bader and M.J.Koczak: European Sci. News & Info. Bul., 1991, vol. 18, pp. 18-24.
2. A.Mortensen and M.J.Koczak: JOM, 1993, vol. 3, pp. 10-18.
3. H.C.Yuen, B.Ralph and W.B.Lee: Scripta Metall. et. Mat., 1993, vol. 29, pp. 695-700.
4. J.C.Lee and K.N.Subramanian: J. of Mat. Sci., 1993, vol. 28, pp. 1578-1584.
5. D.H.Allen and D.C.Lagoudas: Damage Mechanicas in Composites, ASME, USA, 1992, pp. 181-90.

TRIBOLOGICAL STUDIES ON ALUMINIUM MATRIX COMPOSITES

B.N. Pramila Bai, R.A. Saravanan and M.K. Surappa[*]

Department of Mechanical Engineering
*Department of Metallurgy
Indian Institute of Science
Bangalore - 560 012
India

ABSTRACT

Sliding wear and erosive wear studies are being carried out on MMC with different matrices and reinforcements. The general theme of these studies is to understand the mechanism of material removal in these processes. All the composites were processed by cast route and further subjected to secondary processing, i.e. hot extrusion.

Sliding wear studies have been carried out under dry conditions using a pin on disc set up. A cast alloy (A356) reinforced with SiC_p (40μm size) and a wrought alloy (Al2024) reinforced with Al_2O_{3p} (40μm size) have been studied over a range of pressures. In the case of A356, reinforcement with SiC_p improves the wear resistance. This has been attributed to the increased resistance to flow imparted by the presence of SiC_p particles at the surface. SiC_p particles also promote the formation of a protective iron rich layer on the surface which keeps the wear process in the mild wear regime. In contrast, the presence of Al_2O_3 particles in Al2024 matrix do not have positive influence on wear resistance in the as extruded condition, but in T6 condition, the wear rates of the base alloy and the composite are comparable. Under T6 condition the base alloy undergoes seizure where as, the composite dose not.

Inorganic Matrix Composites
Edited by M. K. Surappa
The Minerals, Metals & Materials Society, 1996

Erosion studies with a multi particle erosion rig using quartz particles (750μm & 60μm size) against A356 composite reinforced with SiC$_p$ particles reveale that when size of erodent particles were much larger than the reinforcement particle, as extruded composite is slightly inferior to the base alloy. With an addition of 0.4% Mg in A356 alloy the erosion rate of the composite and of the base alloy were comparable. For impact with fine particles the erosion rate of the composite is either less or comparable to that of the base alloy.

Both sliding wear and erosive studies indicate existence of certain parametric regimes wherein the composite can be a better tribological material than the base alloy. Extensive tribological studies can lead to maximum exploitation of the potential of these new class of materials.

INTRODUCTION

In recent years, MMCs have greatly attracted the researchers in the area of tribological applications. Higher specific modulus and strength values of MMCs compared to the monolithic alloys make them an attractive, cost effective and alternative material to the conventional metallic materials, especially in applications requiring minimum weight for the structures. Recent investigations have proved that many of the MMCs also possess excellent wear resistance. These high wear resistant class of new materials i.e. composites are now being recommended for a number of applications such as bearing sleeves, pistons, cylinder liners etc., where superior wear resistance is called for.. The aim of this paper is to explore the response of the MMCs with different matrices and reinforcements to sliding wear and erosive wear.

In general, for sliding against metals and abrasives, many studies report [1-17] that, composites exhibit better wear resistance than the unreinforced alloy. Further, Axen et al.,[14] noted that, in a variety of wear conditions, the particulate reinforced composites perform better than the fibre reinforced composites. Alphas and Zhang [13], in their work on A356-SiC$_p$ composites sliding against steel in a pin on ring arrangement report that, SiC$_p$ reinforcement plays a beneficial role in improving the sliding wear resistance. Also, Pan et al., [7] in their work on sliding wear in block on ring experiment, have shown that, heat treating to T6 condition, the wear rate of Al2124-SiC$_p$ composite to be less than the as-extruded condition. The over aged specimen shows least wear and minimum damage of the steel ring.

Published results on sliding wear of Al$_2$O$_3$ reinforced composite are varied and often contridicting. A powder metallurgy processed alumina particle reinforced Al-10% Zn composite showed an improvement in wear resistance by two order of magnitude at an optimum content of 30% alumina [14]. In another study, alumina reinforced copper alloys show a linear decrease in wear resistance with increasing particle content

[15]. In case of Al_2O_3 reinforcement, fracture and re-embedding of alumina particles have been reported [4,15].

Erosion by hard particles impact such as SiC or Al_2O_3 have shown poor erosion resistance in the case of composites [16, 18-22]. In a work on erosion study on metal matrix composites by silica, Ninham et al., [18] have reported that, a significant increase in erosion resistance when metals with carbide volume fractions was greater than 80%. According to Hutchings [21], the performance of the composites under erosion environment depends on whether the reinforcement particle/fibre undergoes fracture or not and that the reinforcement particle can be expected to play a positive role when the fracture does not occur.

In slurry erosion experiments employing silica earlier studies have shown that composite material experienced lower material loss than the base alloy. However, Turenne et al., [23], have shown the importance of erodent particle size on slurry erosion. Coarse silica particle cause breakage of reinforcing particles, whereas finer ones deviate or deflected near the surface causing less damage to the reinforcement particle, thus protecting the matrix.

The objective of the present paper is to understand the mechanism of material removal under dry sliding wear of Al2024-Al_2O_3 and A356-SiC_p composites and erosive wear of A356-SiC_p composites. The influence of matrix material, reinforcement and other parameters on wear behaviour is studied. The wear mechanism are identified by scanning electron microscopy, X-ray and EDAX analysis.

EXPERIMENTAL PROCEDURE

MATERIALS

Al_2O_3 and SiC particles of average sizes $18\mu m$ and $40\mu m$ respectively were used as reinforcements. Al_2O_3 particles were dispersed in Al2024 alloy and SiC particles in A356 alloy. The details of the material investigated is given the table.I. All the composite materials were processed by melt stir technique [24]. The cast billets were hot extruded and the tests were done in the extruded condition. The microstructure of the alloys and the composites tested are given the Figs. 1(a-d).

WEAR TEST

Dry sliding wear tests were performed in a pin on disk apparatus. In the case of A356 alloy and its composites, stepped cylindrical pin with 2mm diameter at the rubbing end were used as test specimen. The sliding speed and sliding distance in these experiments were kept constant at 0.5m/s and 450m respectively and bearing pressure was varied from 2 to 26 MPa. In the case of Al_2O_3 reinforced composites, the experiments were carried out using 8mm cylindrical pin. In this experiment a step loading procedure was adopted at a constant speed of 1.5m/s. At each load the pin was slid for 1 min and the incremental load used was 10N. The hardness of the base alloy and the composites are listed in table. II. The as extruded Al2024 alloy and the composite specimens were solutionized at 490°C for 2 hours followed by artificial aging at 215°C for 4 hours and 1 hour respectively.

Erosion tests were conducted using a rotating arm erosion rig similar to that used by Ravikiran et al., [25]. The details of the material used and their mechanical properties are given in the table III & IV.

Quartz particles of average size 60μm and 750μm were used as erodent particles. (These are henceforth referred as fine and coarse particles respectively.) Test were performed at three velocities (18m/s, 28m/s &50m/s) and at three impact angles (15°, 30° & 90°). Cylindrical samples with 15mm in diameter and 10mm in height were used as test specimens. To provide an initial standard surface all the samples were abraded with 600 grit emery paper. Erosion was quantified by mass loss measurement with an accuracy of ±0.1mg. The specimen were weighed before and after the experiments after thorough cleaning with acetone in ultrasonic cleaner. A steady state erosion rate (expressed in mg of the test material removed per gm of the impinging particles) was established for each material initially.

Worn surfaces, sections of worn surfaces and debris were observed using scanning electron microscopy. For sub-surface observation the specimens were given protective nickel coating prior to sectioning and metallographic observation.

RESULTS

1. Dry Sliding Wear

(1a). Wear Results

Figs. 2(a,b & c) shows the sliding wear charteristics of A356 and and Al2024 alloy and composites. It can be observed that SiC reinforced composite (i.e., A356-SiC$_p$) shows positive response whereas Al$_2$O$_3$ reinforced Al2024 composite show negative response in improving wear resistance in the as extruded condition. However, in T6 condition , the Al2024 alloy undergoes seizure whereas the composite does not.

A356-SiC$_p$ composites show less wear than the unreinforced alloy at all pressures and an increase in SiC content decreases wear in the as extruded condition (Fig. 2a). In the case of as extruded Al$_2$O$_3$ reinforced Al2024 composites, from the Fig. 2b, it can be clearly seen that the wear rate of unreinforced alloy is lower than that of the composites. In the T6 condition, (Fig. 2c) upto 30N the wear rate of aged composite was negligible. Beyond 30N and upto 200N, the wear rates of the two materials were comparable. Beyond 200N, the wear rate of the peak-aged unreinforced alloy is slightly higher than that of the composite, and the unreinforced aged alloy eventually seized at 230N. The seizure event was accompanied by a sudden increase in wear rate, heavy noise and vibrations. The peak-aged composite shows an increase in wear rate between 225 and 260N, and beyond that wear rate gradually decreases. However, with further increase in load, both noise and vibration reduced, and the macro seizure encountered in the case of base alloy was not observed in the composites.

(1b). SEM Observation

Scanning electron microscopic observation of the worn surface of A356 alloy shows generally smooth surface with fine grooves. With increasing pressure the shallow craters develops where material has been removed, otherwise surface is smooth. In the case of composite, smooth surfaces and granular rough regions are observed. With increase in SiC content, occurrence of granular rough regions also increases. EDAX analysis of the worn surfaces revealed richer in iron content in the granular region than in the smooth region. Iron content in these regions increased with increasing SiC. Sub-surface observation revealed a clear material flow in the unreinforced alloy (Figs. 3a & b) which was absent in the case of composites.

SEM observation on the worn out surfaces of Al2024 alloy and the composite showed similar features regardless of the load except for the seized condition corresponding to the aged unreinforced alloy. The worn surfaces consisted of long smooth patches with interspersed craters. In addition, in the case of composites, severe localized cracking was also observed. Most of the worn surface of the peak-aged unreinforced alloy shows evidence of severe shearing (Fig. 4a). However, the worn surface of the peak aged composite at 280N revealed absence of any significant shear (Fig. 4b).

EDAX analysis on the worn surface revealed presence of iron in both base alloy and the composites. The iron content in composite at low loads is higher than that present in unreinforced alloy, while at higher loads both are comparable. An important observation is that , peak-aged and seized unreinforced alloy specimen showed absolutely no iron, while that of peak-aged composites worn at 280N showed presence of iron. Sub-surface scanning electron microscopic observation in the case of as-extruded composite reveals fragmentations and dispersion of Al_2O_3 in the near sub-surface region (Fig. 5). The sub-surface of as-extruded base alloy shows bending of extrusion flow lines in the sliding direction.

2. Erosive Wear

(2a). Wear Results

The steady state erosion rate of the base alloy and the composite under coarse particle erosion study are of the same order. Typical plots of the erosion rate vs. impact angle are shown in Figs. 6(a& b).

% increase in erosion rate is calculated as follows for all the material and the data is presented in the table. V.

% increase in E.R. $= \left[\frac{(E_c - E_b)}{E_c} \times 100\right]$

where, E.R is erosion rate and E_c and E_b are steady state erosion rate of composite and unreinforced alloy respectively.

Table. V shows the % change in erosion rate data for all the experimental conditions. A negative value indicates that erosion rate of the composite is less than that of the base alloy. The data shows that, at low impact velocity and at normal incidence, the erosion rates of the composite are in general high. At highest velocity tested (50m/s) and at shallow angle, the erosion rate of the composites are within 13% for the materials A1 and A1C1 in the extruded condition. In the case of the materials A2 and A2C1, the erosion rate of the composite is less than the base alloy under certain conditions.

With fine particle impact, the erosion rates of the composites are in general either comparable or less than that of the unreinforced alloy. Figs. (7a & b) show the representative plot of erosion rate vs. impact angle.

(2b). SEM Observation

Figs. 8(a-d) shows the SEM image of the eroded surfaces of the base alloy and the composite studied under coarse particle impingement. In general the surface of the unreinforced alloy and the composite appear similar. Figs. (8a & b) show low magnification view of the two materials eroded under identical conditions. Extensive plastic deformation with characteristics deep dents, folds and lips are observed. In addition, in the case of composites, regions where the reinforcement particle are almost completely dislodged are seen. Fig. 8c shows a spot where in fragments of SiC are retained but most of the particles are dislodged. In contrast, in the case of A2C1, the retention of the SiC particles is higher, and Fig. 8d shows such a region where SiC particles still embedded in the matrix. Even in the case of shallow angle impact the surface features of the base alloy and that of the composite are similar at low magnification (Figs. 9a & b). At higher magnification, the composite surface exhibits larger number of fine cracks than that observed in the case of base alloy (Figs.

9c & d). In the case of A2C1, the SiC particles appear to resist the gouging process. Fig. 9e shows a SiC particle at the end of a gouged patch suggesting that further gouging might have been resisted by the particles. Another feature common to the both base alloy and the composite is the presence of regions which are fluorescing in SEM. They are associated with angular particles/fragments and EDAX showed them to be rich in Si. A region corresponding to that is shown in Fig. 10. As SiC particles show different morphology and do not fluoresce, these were suspected to the particles of SiO_2 and this was confirmed later.

The general appearance of the base alloy and the composite under fine particle impingement are again broadly similar (Figs. 11a & b). The surface appear smooth and the craters formed are relatively shallow and the formation of the lips around the craters suggests deformation. The main feature observed in the case of the composite is the retention of the reinforcement particle on the surface. Fig. 11c reveals SiC particles which have developed steps due to chipping or small scale fragmentation rather a complete dislodgment of SiC as observed in the case of coarse erodent particle study.

X-ray diffraction studies on the eroded surfaces of the base alloy and the composite reveals extensive silica embedment. Qualitatively, the intensity peaks shows, at low impact velocity embedment of silica was more than at high impact velocity. X-Ray diffraction patterns for the A1C1 composite tested at 18m/s and 50m/s impact velocity and $90°$ impact angle are shown in Figs. (12a&b).

Optical microscopic observation of sub-surface reveal extensive plastic deformation in the near surface below the sites of the impact damage. Compared to unreinforced alloy, in the case of A1C1 and A2C1, presence of extensive shear localization is more (Figs.13a,b & c). SEM observation on sub-surfaces reveals presence of cracks in the SiC particles Fig. 14. This cracking could lead to the dislodging of SiC particles.

DISCUSSION

Sliding wear

The present study on sliding wear on A356 and Al2024 alloy reinforced with SiC and Al_2O_3 particles respectively shows two different behaviour. In the case of A356, reinforcement with SiC particles improves the wear resistance. This corroborates well with earlier observation on Aluminium alloy SiC composites. Improvement in sliding resistance has been attributed to the increased resistance to flow imparted by the presence of SiC particles at the surface. SiC particles also promote the formation of protective iron rich layer on the surface which keeps the wear process in the mild regime.

In contrast, the presence of Al_2O_3 particles in Al2024 matrix do not show any positive influence on wear resistance in the as extruded condition. Unlike SiC particles which are known to bear the load and prevent the surface plastic flow [13], the Al_2O_3 particles fragment and gets dispersed [14]. Thus the fragmented Al_2O_3 particles leads to weaker interface, resulting in more discontinuous composite leading to higher wear rate. But, in T6 condition, at low loads, the wear rates of the composite and the unreinforced alloy are comparable. The presence of iron rich layer at lower loads are higher in the case of the composite compared to that of the unreinforced alloy. The presence of iron rich layer can improve wear resistance and this could be the possible reason for the comparable wear resistance at low loads in T6 conditions. However, the peak-aged alloy undergoes seizure whereas the composite does not. The possible explanation for the absence of seizure is related to picking of iron from the counterface. For Al-Si alloys it has been well established that, when iron rich layers are formed the wear is low or mild [26]. In the present case also, the absence and presence of iron rich layer on the worn surfaces of base alloy and the composite supports the above behaviour.

Erosive Wear

In the case of impingement by coarse erodent particle, SEM studies on surface and sub-surface have shown that for A1C1 material, SiC particles are cracked and then more or less completely dislodged. The absence of beneficial effect of SiC particles due to their dislodgment from the matrix clearly indicates the poor performance of the composite. The resistance to fragmentation and dislodging can be improved if the interfacial bond strength of SiC particle-matrix interface is improved. In the case of A2C1, where an excess of magnesium is present, interfacial bond strength is expected to increase and hence the SiC particle are better retained. This is also associated with lower erosion rates in case of A2C1. This corroborates well with the remarks made by Hutchings [21], that, if the reinforcement particles do not fracture the erosion resistance is better.

In the case of composites, the erosion process observed here is associated with adiabatic shear flow. The susceptibility of the composite and the base alloy to undergo adiabatic shear flow under very high strain rate such as ballistic impact [27], as well as under medium strain rate $1S^{-1}$ is already reported [28]. It is also reported that, there is shift in the instability regime towards the lower temperature in the case of the composite at similar conditions. This could be a possible explanation for the absence of any significant adiabatic shear flow in the base alloy. (i.e. the rise in temperature during the erosion process is not sufficient enough for the base alloy to undergo adiabatic shear flow.). This also explains the presence of large amount of fine cracks in the case of composites compared to that of the unreinforced alloy

(Figs. 9c & d). The thickness of the top surface layers to adiabatically deform and fail, probably decide the amount of material removal.

Another interesting and important feature of the coarse particle erosion study is the embedment of SiO_2. In the case of the composite, earlier studies has shown such an embedment in the initial stages. However, for the present material it is the characteristic feature of the steady state process. This indicates the nature of the surface being eroded is dynamically changing. On one hand, an increase in weight of the sample due to embedment should result in lower overall weight loss. On the other hand, higher erosion loss can occur since surface becomes much more discontinuous and the particles are relatively more prone to fracture than the SiC particles. This could also possibly account for the difference in erosion rates between the two materials at different velocities. As mentioned earlier, the SiO_2 embedment is more at low velocity. Thus, even if SiC particles are retained at low velocity and at shallow angle, because of SiO_2 embedment the matrix becomes more heterogeneous and discontinuous resulting in higher erosion rate.

In the case of erosion by fine particles, by and large the composite shows positive role of the reinforcement. Both at normal and shallow incidence, the SiC particle undergoes small scale fragmentation/chipping. This process is a gradual and hence the reinforcement plays a positive role. At any instant as soon as the matrix material is removed, the SiC particles take the brunt of the impingement. SiC particle retention is better at shallow angle and hence, the erosion resistance of the composite enhanced at shallow angle.

The present work suggest that the composite can find application particularly when one is dealing with fine particle impact. Further, with fine erodent particle, the operating erosion mechanism being mild deformation and fragmentation, the beneficial effects of reinforcement can be enhanced by adopting T6 heat treatment for improving the erosion resistance.

CONCLUSION

1. SiC_p reinforcement has a beneficial effect in improving sliding wear resistance of A356 Alloy.

2. The presence of SiC particles in A356 matrix, increases the resistance to flow at the surface and aids in the formation of iron rich layers during sliding, there by improving the wear resistance of the composite.

3. The peak-aged Al204-15 vol.% Al_2O_3 composite shows better seizure resistance than does the unreinforced alloy.

4. In the as-extruded condition, Al2024-15 vol.%Al_2O_3 composite shows poor sliding wear resistance compared to the unreinforced alloy.

5. If erodent particles are coarse, erosion resistance of the composite are inferior or comparable to that of the base alloy and if it is fine, the composite shows better erosion resistance.

6. The features of the coarse particle impact are dislodging, SiO_2 embedment and adiabatic shear flow, while that of fine particle relatively higher degree of SiC particle retention.

7. In general, the erosion resistance of the composite depends on the size of the erodent particle, the matrix chemical composition, impact velocity and impact angle.

ACKNOWLEDGMENTS

The authors wish to thank Department of Science and Technology for the financial support during the course of this study.

REFERENCES

1. A. Sato and R. Mehrabian:*Metall. Trans. B*, 1976, Vol.7, p. 443.
2. M.K. Surappa, S.V. Prasad and P.K. Rohatgi: *Wear*, 1982, Vol. 77, p. 295.
3. F.M. Hosking, F.F.Portillo, R.Wunderlin and R. Mehrabian: *J. Mater. Sci.*, 1982, Vol.17, p. 477.
4. K. Anand and Kishore: *Wear*, 1983, Vol. 85, p. 163.
5. N. Saka and D.P. Karalekas: *Proc. Int. Conf. Wear of Mater.*, Vancouver, 1985, p. 784-793.
6. F. Rana and D.M. Stefanescu: *Metall. Trans. A*, 1989, Vol. 20, p. 1564.
7. A.T. Alphas and J.D Embury: *Scr. Metall. Mater.*, 1990, Vol. 24, p. 931.
8. A. Jokinen and P. Andersson: *Proc. Powder Metallurgy Conf. and Exhib.*, Metal Powder Industries Federations, Pittsburgh, PA, 1990.
9. Y.M. Pan, M.E. Cheng and M.S. Fine: *Scr. Metall. Mater.*, 1990, Vol. 24, p.1341.
10. A.R. Nesarikar, S.N. Tiwari and E.E Graham: *Mater. Sci. Eng.*, A, 1991, Vol. 147, p. 191.
11. M. Roy, B. Vekataraman, V.V. Bhanuprasad, Y.R. Mahajan and D. Sundarajan: *Metall. Trans. A*, 1992, Vol. 23, p. 2833.
12. O.P Modi, B.K. Prasad, A.H. Yagneswaran and M.L. Vaidya: *Mater. Sci. Eng. A*, 1992, Vol. 151, p. 235.
13. A.T. Alphas and J. Zhang: *Wear*, 1992, Vol. 155, p.83.

14. N. Axen, A. Alahelisten and S. Jacobson. Personal communication. 19xx.
15. J. Zhang and A.T. Alphas: *Mater. Sci. Eng.*, A, 1993, Vol. 161, p. 273.
16. A. Alahelisten. F. Bergman. M. Olsson and S. Hogmark: *Wear*. 1993. Vol. 165. P.221.
17. A.T. Alphas and J. Zhang: *Metall. Trans. A*, 1994 Vol. 25, p. 969.
18. A.j. Ninham and A.V. Levy: *Wear*. 1988, Vol. 121, p. 347.
19. Sreeram Srinivasan, R.O. Scattergood and R. Warren: *Metall. Trans. A*. 1988, Vol. 19a, p. 1785.
20. A. Wang and I.M. Hutchings: *Metal and Ceramic Matrix composites: Processing. Modelling and Mechanical Behaviour*, R.B. Bhagat. A.H. Claues. P.Kumar and A.M. Ritter eds., The Minerals, Metals and Material Society, OH, 1990, p. 499-508.
21. I.M. Hutchings: *Proc. of the 2nd Eru. Conf. on Advanced structural materials*, University of Cambridge, Cambridge, UK, Vol.2, 1991, p. 56-64.
22. W. Wu, K.C. Goretta and J.L. Routbort: *Mater. Sci. Eng.*, A, 1992, Vol. 151, p. 85.
23. S. Turenne, Y. Chatigny, D. Simard, S.Craon and J. Masounave: *Wear*, 1990, Vol. 141, p. 147.
24. M. Vasudevan and M.K. Surappa: *Proc. Int. Conf. on Adavnces in Composite Materials, 14-18, January, 1990*, Oxford and IBH Bombay, 1991, p. 265.
25. A. Ravikiran, K.N. Natarajan, S.K. Biswas, and B.N. Pramila Bai: *Wear*, 1990, Vol. 141, p. 85.
26. A. Somi Reddy, B.N. Pramila Bai, K.S.S. Murthy, and S.K. Biswas: *Wear*, 1994, Vol. 115, p. 171.
27. Sunghak Lee, Kyung Mox Cho, Ki Chong Kim and Won Bong: *Metall Trans A*, 1993, Vol. 24a p. 895.
28. J. Sarkar: *M. Sc Thesis*, 1993, I.I.Sc, Bangalore, India.

LIST OF TABLES

I. Materials used for sliding wear experiments.
II. Hardness of the base alloy and composites used for sliding wear experiments.
III. Materials and their codes used for erosion experiments.
IV. Mechanical Properties of the material used for erosion experiments.
V. % increase in erosion rate data for all the material tested for all experimental conditions.

TABLE I

Materials
* Al 2024 alloy
* Al 2024 - 15% Al_2O_3
† A356 alloy
† A356 - 15% SiC_p
† A356 - 25% SiC_p

* Both as-extruded and T6 condition.
† Only as-extruded condition.

TABLE II

Material	Hardness(VHN)
* Al 2024	83
† Al 2024	152
* Al 2024 - 15% Al_2O_3	103
† Al 2024 - 15% Al_2O_3	148
A356 alloy	55
A356 - 15% SiC_p	62
A356 - 25% SiC_p	74

* as-extruded
† as-extruded & T6 heat treated

TABLE III

Si	Mg	Fe	Cu	Ti	Al	%SiC	CODE
7	0.34	0.14	0.01	0.01	balance	-	A1
6.9	0.71	0.12	0.01	0.01	balance	-	A2
6.9	0.3	0.15	0.01	0.01	balance	10	A1C1
6.9	0.68	0.15	0.01	0.01	balance	10	A2C1

TABLE IV

Material	Hardness, VHN	Youngs Modulus, GPa	UTS, MPa	% elongation
A1	48.1	67.41	199.64	24
A1C1	65.3	73.93	180.5	14
A2	53.3	70.63	230.7	20
A2C1	70.2	75.11	222.5	12

TABLE V

Imapact Velocity m/s	% Increase in Erosion Rate (for A1 and A1C1)					
	Coarse Particle			Fine Particle		
	15°	30°	90°	15°	30°	90°
18	19	16	30	10	0	4
28	9	17	15	-4	-9	-2
50	13	9	6	3	-6	13

Imapact Velocity m/s	% Increase in Erosion Rate (for A2 and A2C1)					
	Coarse Particle			Fine Particle		
	15°	30°	90°	15°	30°	90°
18	4	6	14	-8	-30	--
28	17	6	3	8	10	10
50	12	-13	-3	-6	-10	-5

% Increase in Erosion Rate $= \left[\frac{(E_c - E_b)}{E_c} \times 100\right]$

where, E_c and E_b are the erosion rates of composite and base alloy respectively.

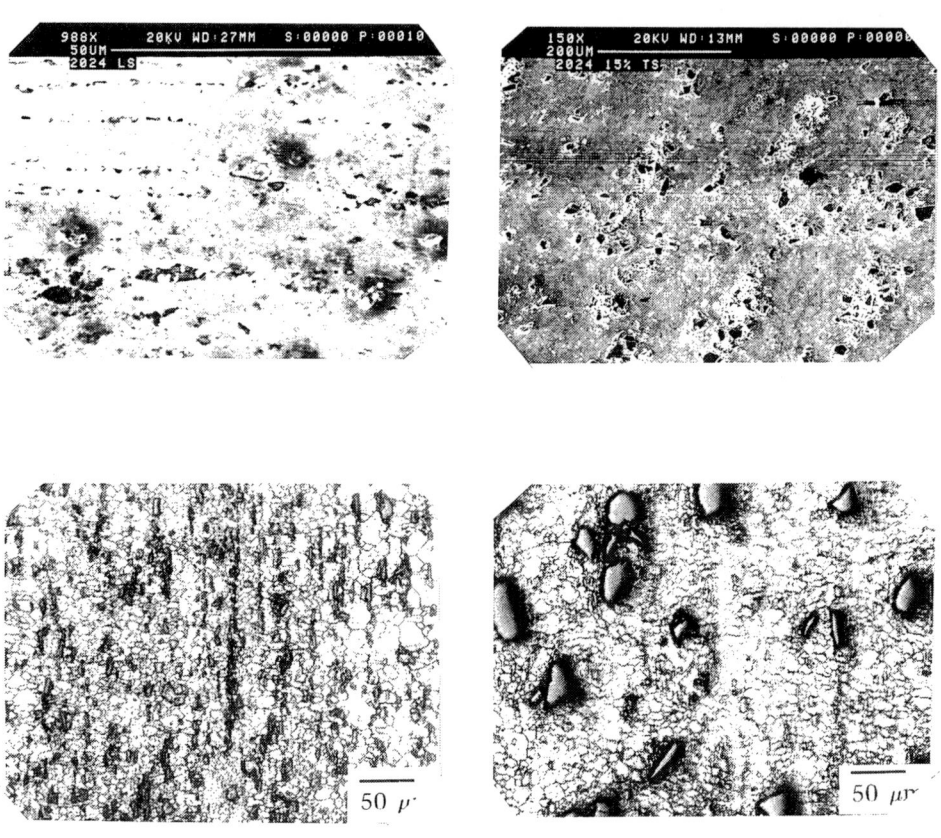

Figs. 1(a-d) Optical micrographs of
(a) Al2024 alloy; (b) Al2024 – 15% Al$_2$O$_3$; (c) A356; (d) A356 – 10%

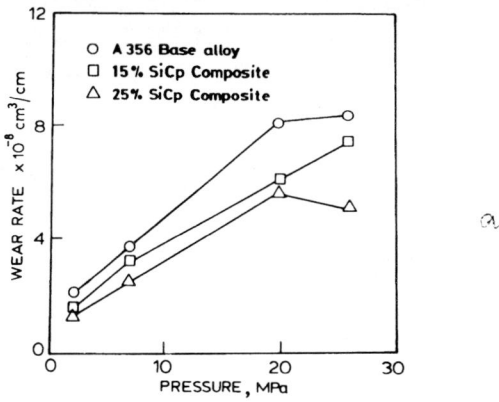

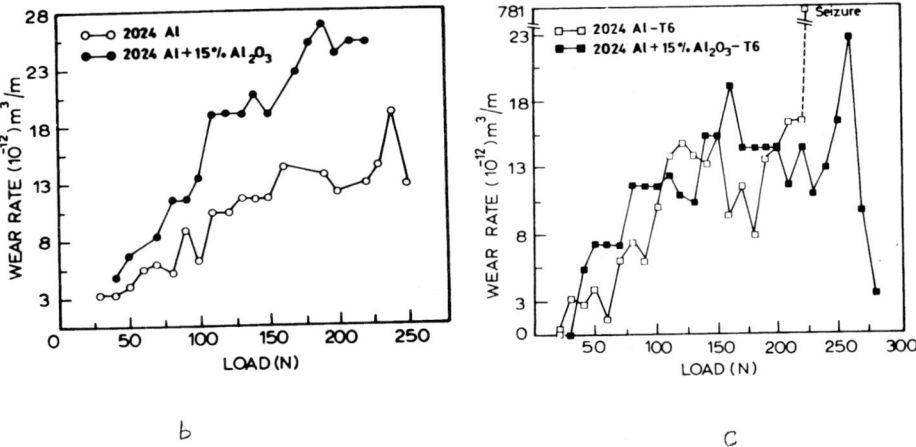

Figs. 2(a-c) (a). Variation of rate with pressure for A356 and A356 - SiC_p.
(b). Comparison of wear rates as a function of load for the as-extruded Al 2024 alloy and Al2024 - 15%Al_2O_3 composite.
(c). Comparison of wear rates as a function of load for the peak aged Al alloy 2024 and Al2024 - 15%Al_2O_3 composite.

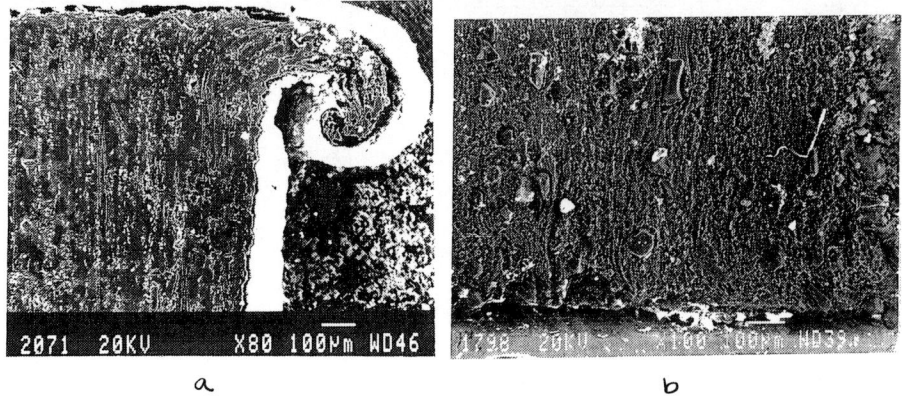

Figs. 3(a-b) SEM images of worn sub-surfaces of A356 unrinforced alloy and SiC$_p$ reinforced composite.
(a). Presence of flow on the surface in the case of A356 unreinforced alloy. (b). Absence of material flow in the case of A356-SiC$_p$ composite.

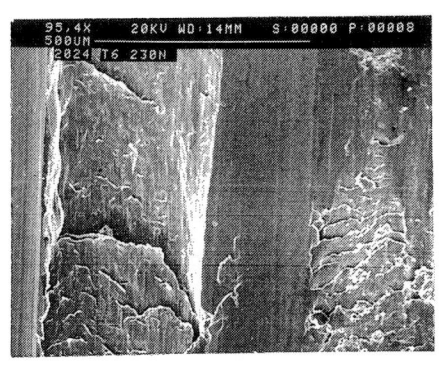 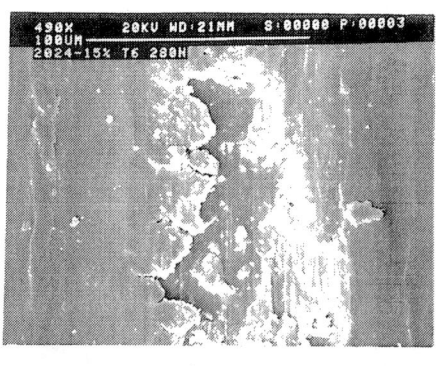

Figs. 4(a-b) SEM images of worn surfaces of Al2024 unrinforced alloy and Al2024 - 15%Al$_2$O$_3$ reinforced composite in the peakaged condition.
(a). Presence of severe shearing in the unreinforced alloy at 230N.
(b). Absence of shearing in the case of composite at 280N.

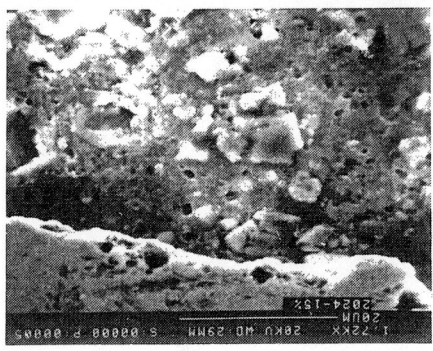

Fig. 5 SEM Micrograph revealing fragments and dispersion of Al_2O_3 in the near sub-surface.

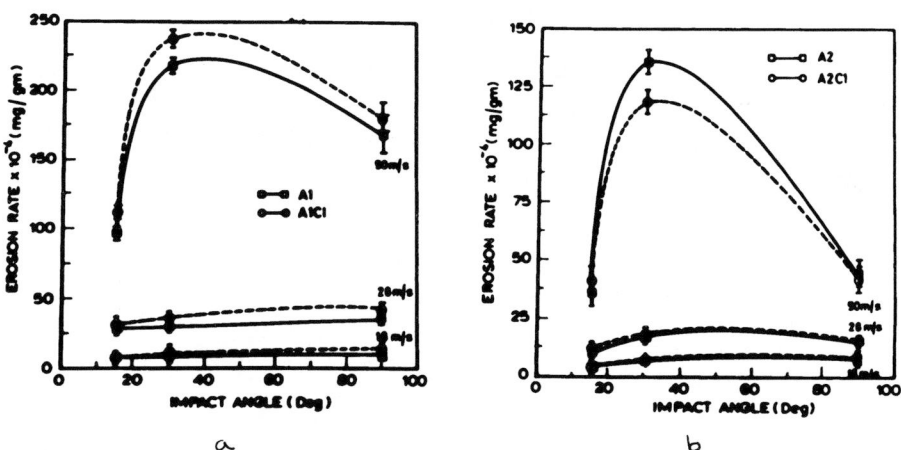

Figs. 6(a-b) Erosion rate (expressed as mass loss in mg per unit mass in gm of erodent particle) plotted against impact angle tested with coarse erodent particle for the material. (a) A1, A1C1; (b) A2, A2C1.

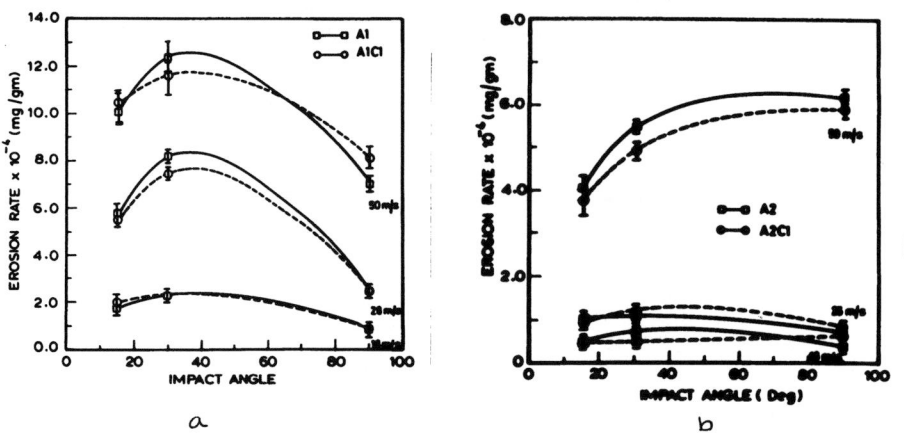

Figs. 7(a-b) Erosion rate (expressed as mass loss per unit mass of erodent particle) plotted against impact angle tested with fine erodent particle for the material, (a)A1, A1C1; (b) A2, A2C1.

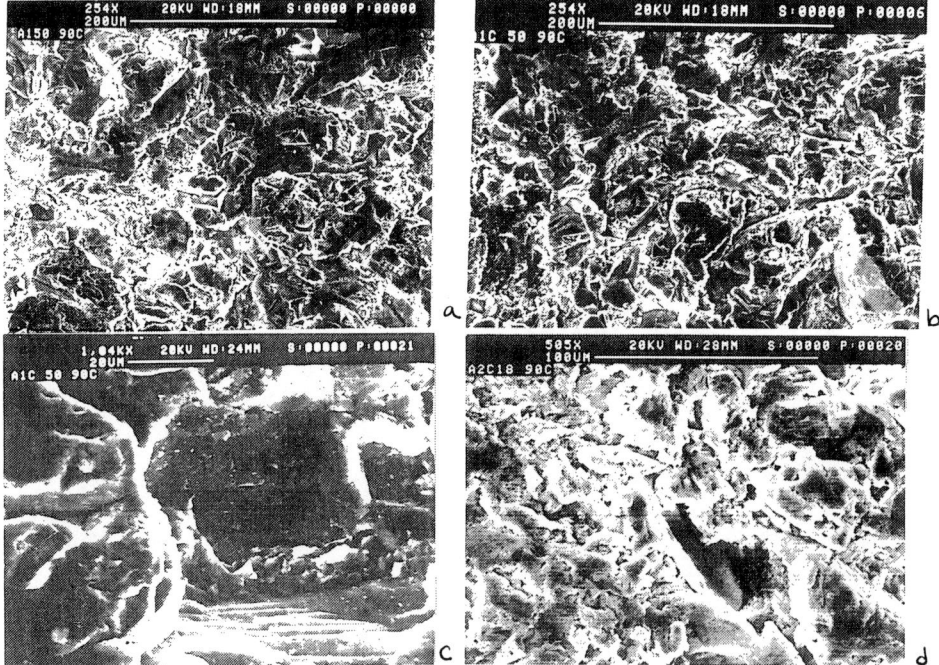

Figs. 8(a-d) SEM micrographs of eroded surfaces when impinged with coarse particle
(a) Low magnification image of A1 impacted at 50m/s and at 90° impact angle.
(b) Low magnification image of A1C1 impacted at 50m/s and at 90° impact angle.
(c) High magnification image of fig.b showing the region where SiC particle is removed.
(d) High magnification image of the region of A2C1 impacted at 18m/s and at 90° showing where the SiC particle is retained.

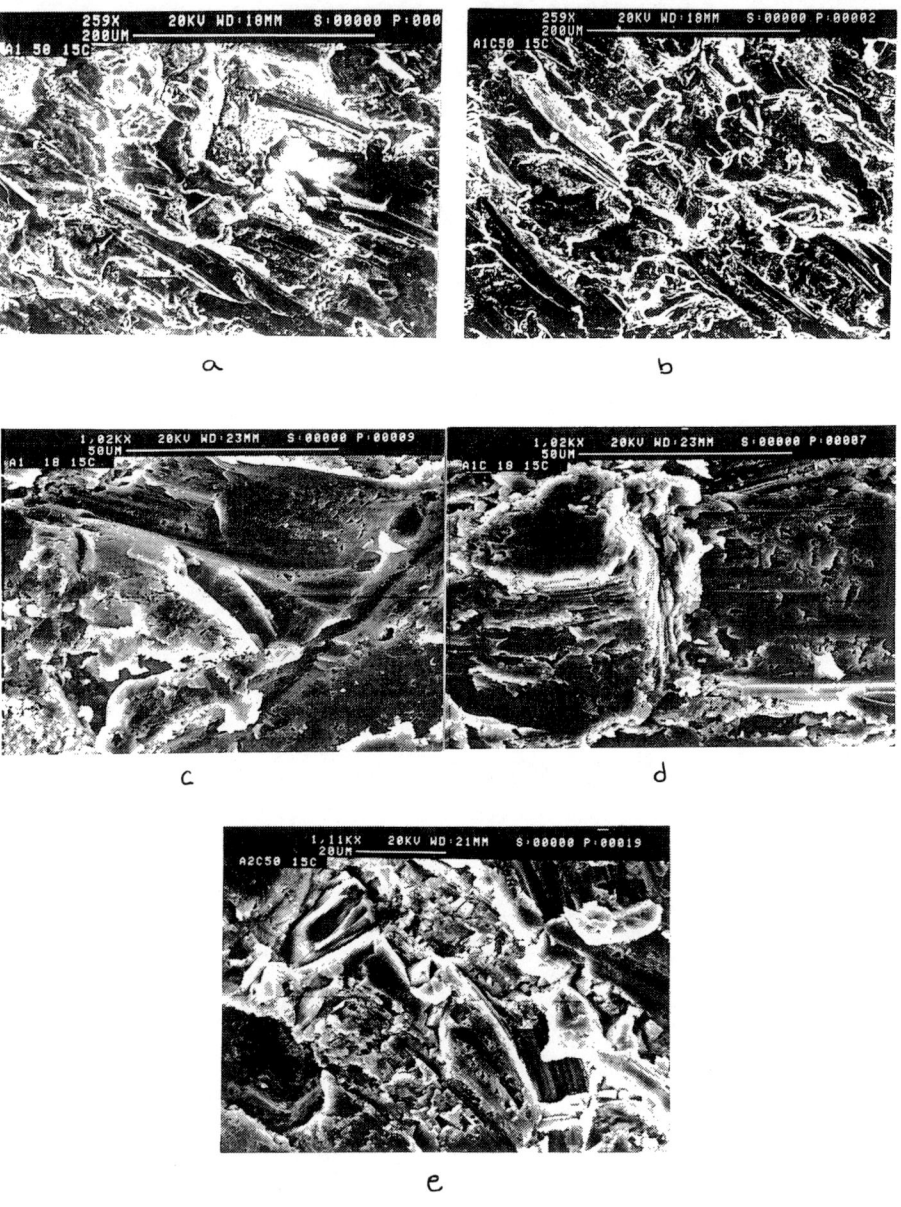

Figs. 9(a-e) SEM micrographs of eroded surfaces when impinged with coarse particle; (a) Low magnification image of A1 impacted at 50m/s and at 15° impact angle. (b) Low magnification image of A1C1 impacted at 50m/s and at 15° impact angle. (c) High magnification image of A1 showing presence of less number of fine cracks. (d) High magnification image of A1C1 showing presence of more number of fine cracks. (e) High magnification image of the region where the SiC particles resist the gouging process in A2C1 material at 50m/s and at 15° impact angle.

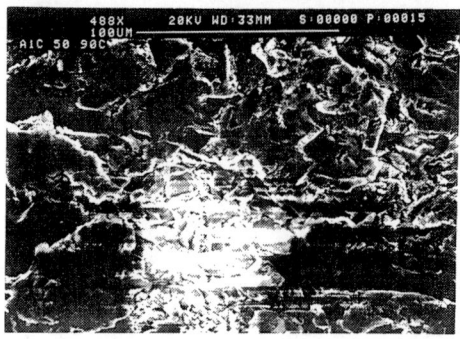

Fig. 10 SEM image showing the region of silica embedment in A1Cl material impacted at 50m/s and at 90° impact angle.

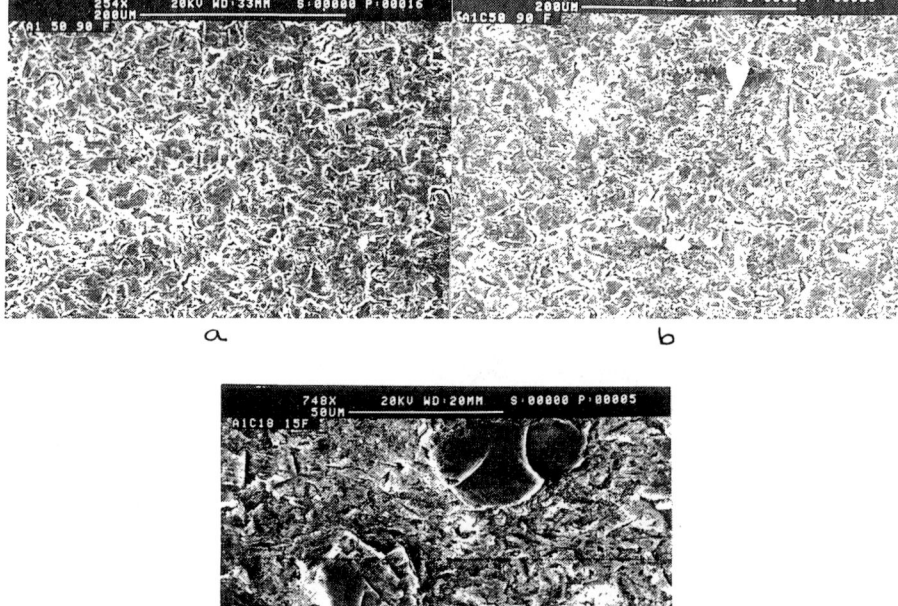

Figs. 11(a-c) SEM micrograph of eroded surfaces corresponding to fine particle impingement
(a) Low magnification image of A1 material at 50m/s and 90° impact angle
(b) Low magnification image of A1C1 material at 50m/s and 90° impact angle
(c) High magnification image of the region showing small scale fragmentation in A1C1 at 18m/s and at 15° impact angle.

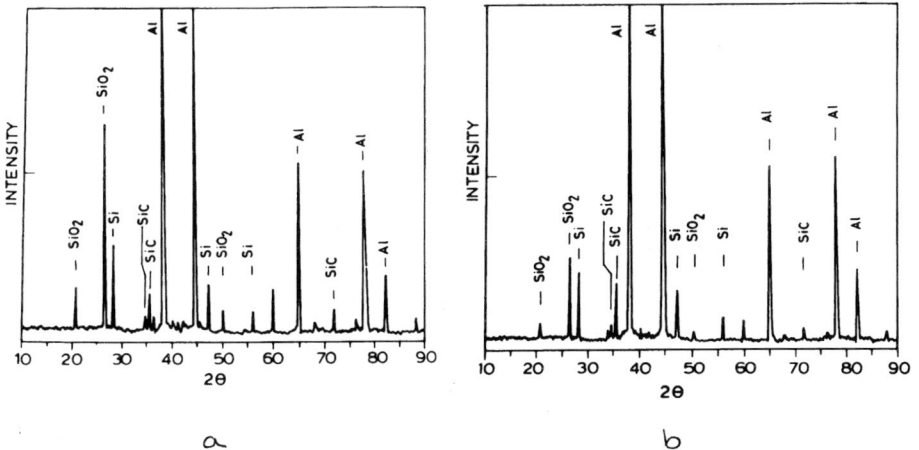

Figs. 12(a-b) X-ray diffraction pattern on the surface of A1C1;
(a) After the erosion test conducted at 18m/s and at 90° impact angle,
(b) After the erosion test conducted at 50m/s and at 90° impact angle.

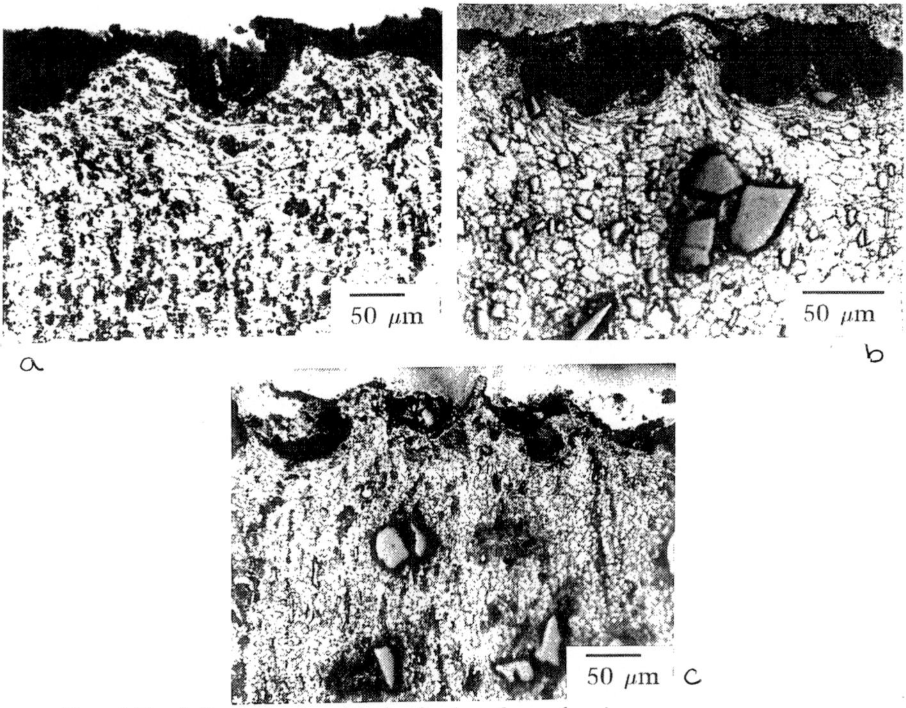

Figs. 13(a-c) Optical micrograph of subsurfaces showing;
(a) Extensive material flow in A1 at 50m/s and at 90° angle,
(b) Extensive material flow in A1C1 at 50m/s and at 90° angle with characteristics shear flow shown by arrow mark in the figure,
(c) Region showing similar features as seen in Fig.b in the material A2C1 under identical condtions.

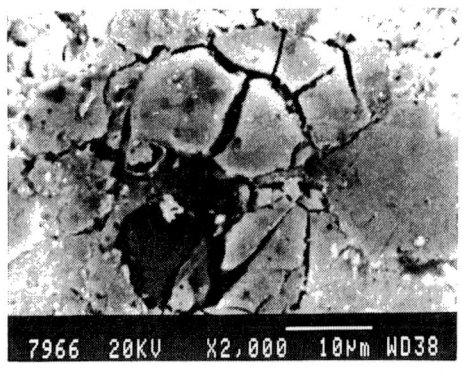

Fig. 14 SEM micrograph showing SiC fragmentation in the subsurface for the material A1C1 tested at 50m/s and 90° impact angle.

INFILTRATION OF Al_2O_3 - Al COMPOSITES INTO COATED SILICON CARBIDE

V.Jayaram[*], Sandeep Kumar[*], T.V.Mani[•], M.S.M.Saifullah[*], J.Sarkar[*], and K.G.K.Warrier[•]

[*] Department of Metallurgy, Indian Institute of Science, Bangalore 560 012, INDIA.

[•] Regional Research Laboratory, Trivandrum 695 019, INDIA.

ABSTRACT

Aluminium alloys have been oxidatively infiltrated into SiC particulate that are coated with SiO_2 or Al_2O_3. The influence of the coating has been examined on the silicon in the alloy channels of the composite, the residual silicon carbide content and the corrosion behaviour of the composite in moisture.

Introduction

Directed melt oxidation [1-7] and infiltration of aluminium alloys into SiC [8,9] pose interesting issues and challenges. Oxidation of SiC to SiO_2 and its subsequent reduction by Al to elemental silicon can result in, (i) enhancement of the Si content of the residual alloy channels, (ii) changes in infiltration rate and microstructure owing to both, increased wettability between particle and melt as well as due to the need to diffuse Si away from the growth surface, and (iii) in extreme cases of oxidation, a substantial loss of SiC. In addition, reaction between Al and SiC to form Al_4C_3 is avoided by including Si in the alloy reservoir and which remains (with some further enrichment) in the composite alloy channels as a hard and brittle constituent.

Flexibility in tailoring the final composite requires that the reactions between SiC and oxygen/aluminium be eliminated or at least controlled. For example, fine particulate ($\leq$ 5μm) are not normally retained owing to extensive oxidation during the long exposures to the oxidising ambient. Then again, the composition and amount of residual alloy can significantly alter mechanical properties such as sliding wear resistance [10].

In the present work, attempts were made to examine two aspects of the above. Firstly, pre-oxidation of silicon carbide was carried out prior to infiltration to examine the consequences on the microstructure, in particular on the residual silicon content of the alloy channels. In the other set of experiments coatings of alumina were applied on SiC by wet chemical methods. These coated particulate were then infiltrated with Si-containing alloys as well as Si-free alloys and the resulting composites were analyzed to determine the extent of reaction between the particulate and the alloy/atmosphere, during infiltration.

Experimental Procedure

For pre-oxidation studies SiC with a mean particle size of 50µm was pressed with binder into cylindrical pellets of 7mm dia and then heated at 1400°C for times of 0-16 hours, to generate weight changes of 0 to 7.5%. These preforms were infiltrated to ~5mm thickness with Al-10Si-6Zn-0.25Mg at 1100°C in air. Alumina coatings were applied by two methods. In the first case, SiC particles (mean size of 50µm) were sensitized at 80°C and ultrasonically dispersed in a 2 wt.% boehmite sol. After filtration the sol coated SiC was dispersed in n-butanol and spray dried with inlet and outlet temperatures of 180°C and 100°C, respectively. In the second method of coating [11] SiC of mean particle size of 5.6µm was sedimented for 2 hours to yield a suspension with particles no greater than ~5µm. These were then dispersed in aluminium sulphate solution (47g/l) which was heated to 90°C after which urea (52g/l) was added to initiate the reaction which yielded aluminium hydroxide. The coated particles were dried and calcined at 1000°C for 5 hours to produce α-Al_2O_3 coatings. Similar coatings were made on a small batch of β-SiC whiskers to obtain an idea of the thickness of alumina.

Infiltration of Al_2O_3/Al composite into spray coated SiC was carried out at 1000 and 1200°C with a silicon free alloy,

Al-5Zn-0.3Mg. Pieces of composite were then cut and placed in a beaker that was exposed to 100% humidity at $25^{\circ}C$ and the weight of the sample was periodically monitored to check for evidence of moisture degradation. Infiltration was carried out into precipitation coated fine SiC with Al-10Si-6Zn-0.25Mg at $1000-1100^{\circ}C$ in air.

All composites were sectioned for metallographic examination. Volume fractions of the various phases were determined by optical image analysis in the case of the 50μm particulate. Free aluminium was determined by atomic absorption after leaching the crushed composite in HCl.

RESULTS & DISCUSSION
Pre-Oxidation

The oxidation of SiC preforms at $1400^{\circ}C$ is shown in fig.1 and is seen to be parabolic with time. The volume fractions of the various phases and the hardness in the final composite are summarised in table 1. It is noteworthy that the residual silicon content does not increase monotonically with pre-oxidation time despite the enrichment in Si over the starting value of 10%. In fact, oxidation of SiC during the infiltration itself appears to be able to contribute significantly as indicated by sample A. Porosity content does increase with silica content, a feature which is similar to the trend in porosity noted previously with diminishing particle size and may be attributed to the tendency for liquid alloy to climb rapidly along the wetting interface with silica before oxidative growth to fill the pore is complete. Hardness does increase with silicon content though the amount of variation is less than 10%

SPRAY PYROLYSED ALUMINA COATINGS

The composites prepared by infiltrating Al_2O_3/Al from Al-5Zn-0.3Mg, all indicate presence of silicon (table 2), thereby suggesting that the spray pyrolysed Al_2O_3 did not afford complete protection. The result of degradation tests in 100% humidity are shown in fig.2. The noteworthy results are

(i) that the uncoated particulate suffer greater degradation after infiltration at 1000^oC compared to 1200^oC while for coated particulate there is little change in degradation with process temperature.

(ii) that the coatings plays a substantial note in reducing degradation at 1000^oC and not at 1200^oC and

(iii) that the least silicon is found after infiltration at 1000^oC with coated particulate. All the weight gains in figure 3 are attributed to hydrolysis of Al_4O_3 according to $Al_2O_3 + 12H_2O \longrightarrow 4Al(OH)_3 + 3CH_4$. These results strongly suggest that while the coating plays some role during infiltration at lower temperatures, there is equal benefit to be derived at higher temperatures by a partially protective layer of silica that would in any case be produced by thermal oxidation of SiO_2. This layer of silica could delay contact between particle and melt and also serve to pump silicon into the alloy by displacement reaction, both of which would reduce the formation of Al_4C_3.

ALUMINA COATINGS BY PRECIPITATION

Typical coating thicknesses/morphology re shown on SiC Whiskers in fig.3. On 5μm particulate SiC the coatings showed no evidence of cracks or discontinuities. After infiltration at $1000^\circ C$ a high volume fraction of particulate could be retained with little porosity as shown in fig.4, in sharp contrast to the extensive reduction in SiC content and the increase in porosity that has been observed earlier with uncoated particulate [9,10]. This last technique therefore appears promising as a route to the fabrication of high volume fraction, fine SiC reinforced Al_2O_3/Al composites.

Acknowledgements

Financial support for this work was made possible by a grant from the Aeronautical Research and Development Board, Government of India.

FIGURE CAPTIONS

1. Weight change, W, during oxidation of SiC (mean particle size of 50µm) at 1400°C.
2. Weight change due to degradation of composites made with Si-free alloys in 100% humidity at room temperature.
3. Cracked coatings of alumina on SiC Whiskers produced by precipitation.
4. Comparison of microstructure produced after infiltration into coated (top) and uncoated (bottom) SiC (mean particle size 5µm). Coatings result in greater retention of SiC, reduced porosity and less alloy phase (bright regions in lower micrograph).

REFERENCES

1. M.S. Newkirk, A.W. Urquhart, H.R. Zwicker, and E. Breval, *J. Mater. Res.* **1**, 81-89 (1986).
2. M.S. Newkirk, H.D. Lesher, D.R. White, C.R. Kennedy, A.W. Urquhart, and T.D. Claar, *Ceram, Eng. Sci.* Proc. **8**, 879-885 (1987).
3. O. Salas, H. Ni, V. Jayaram, K.C. Vlach, C.G. Levi and R. Mehrabian, *J. Mater. Res.* **6**, 1964 (1991).
4. A.S. Nagelberg, *Solid St. Ionics.* **32/33**, 783 (1989).
5. A.S. Nagelberg, S. Antolin and A.W. Urquhart, *J. Am. Ceram. Soc.* **75**, 447 (1992).
6. A.S. Nagelberg, S. Antolin and A.W. Urquhart, *J. Am. Ceram. Soc.* **75**, 455 (1992).
7. O. Salas, V. Jayaram, K.C. Vlach, C.G. Levi and R. Mehrabian, in *Processing and Fabrication of Advanced Materials for High Temperature Applications* (Edited by V.A. Ravi and T.S. Srivatsan), p.143 TMS, Materials Park, Ohio (1992).
8. E. Manor, H. Ni, C.G. Levi and R. Mehrabian, *J. Am. Ceram. Soc.* **73**, 2615 (1990).
9. S.P. Dhandapani, V. Jayaram and M.K. Surappa, *Acta. Metall. Mater.* **42.** 649-656 (1994).
10. V. Jayaram, R. Manna, K. Kshetrapal, J. Sarkar and S.K. Biswas, Submitted to *J. Am. Ceram. Soc.*
11. H. Nakamura, and A. Kato, Ceramic International, **18**, 201 (1987).

TABLE 1

VOLUME FRACTIONS OF PHASES IN COMPOSITES PRODUCED AFTER PRE OXIDATION

OXIDATION TIME (hrs)	0	4	8	16
SiC	56.2	47.2	45.8	49.8
Al_2O_3	26.6	31.7	35.5	29.8
VOIDS	3.1	7.0	7.5	8.2
Al*	7.5	5.7	6.6	6.5
Si*	6.6	8.4	4.6	5.7

* ± 1%

TABLE 2

VOLUME FRACTIONS OF PHASES IN COMPOSITES PRODUCED WITH COATED/UNCOATED SiC

	1000°C		1200°C	
	COATED	UNCOATED	COATED	UNCOATED
SiC[1]	24.1	44.6	28.3	45.5
Al_2O_3[1]	62.4	41.6	50.7	37.3
VOIDS[1]	4.0	4.1	9.0	7.6
TOTAL ALLOY[1]	9.5	9.7	12.0	9.6
Al[2]	8.6	7.9	10.5	7.8
Si[3]	0.9	1.8	1.5	1.8

1. OPTICAL IMAGE ANALYSIS
2. ATOMIC ABSORPTION
3. DIFFERENCE BETWEEN TOTAL ALLOY AND Al. PRECISION IN Si ~ ± 5% TOTAL ALLOY, i.e., ± 0.5%.

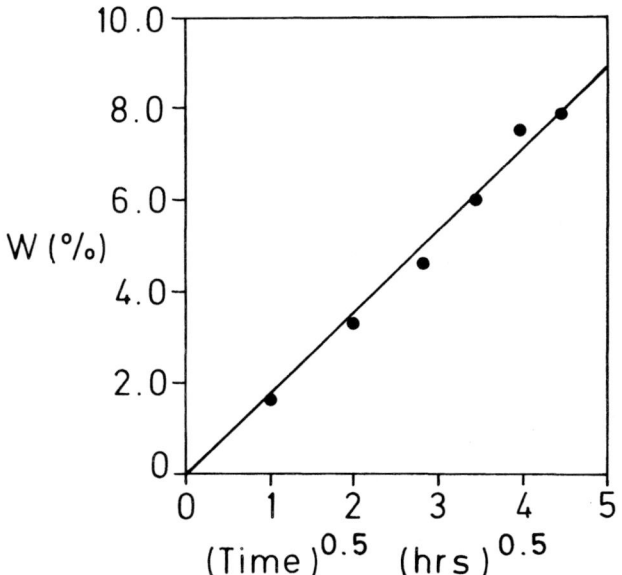

Fig.1. Weight change, W, during oxidation of SiC (mean particle size of 50μm) at 1400°C.

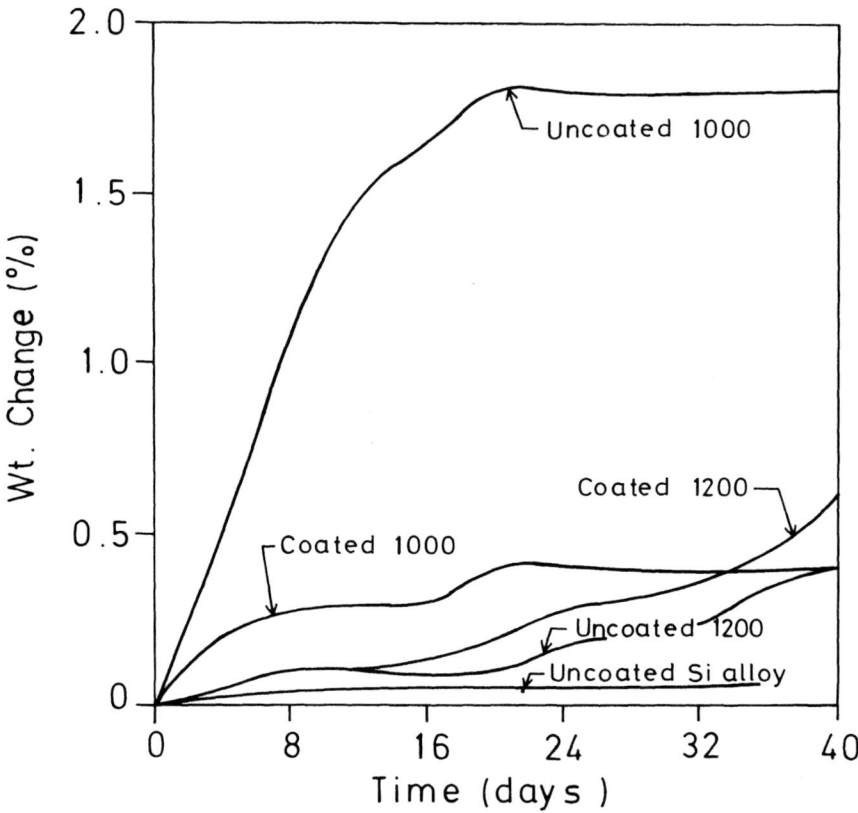

Fig. 2. Weight change due to degradation of composites made with Si-free alloys in 100% humidity at room temperature

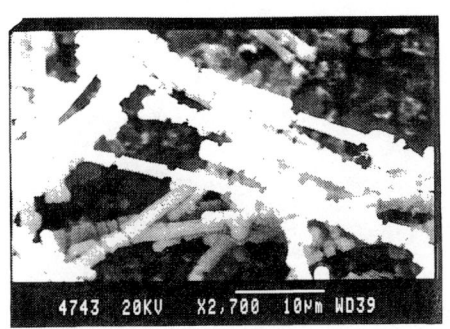

Fig.3. Cracked coatings of alumina on SiC Whiskers produced by precipitation.

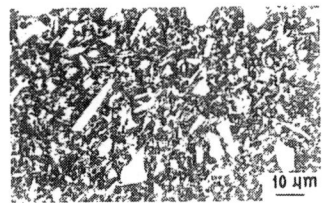

Al$_2$O$_3$ coated SiC$_p$ (5 μm)

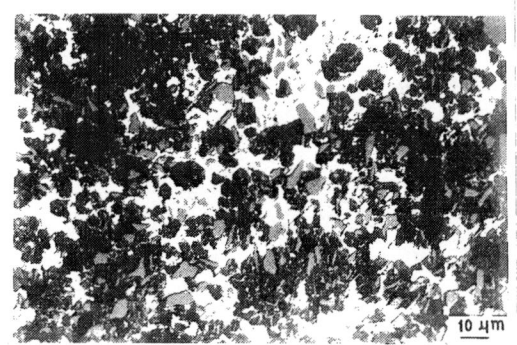

Uncoated SiC$_p$ (5 μm)

Fig.4. Comparison of microstructure produced after infiltration into coated (top) and uncoated (bottom) SiC (mean particle size 5μm). Coatings result in greater retention of SiC, reduced porosity and less alloy phase (bright regions in lower micrograph).

COMBUSTION SYNTHESIS OF MoSi$_2$ BASED COMPOSITES

J. SUBRAHMANYAM

Combustion Synthesis Group, Defence Metallurgical Research Laboratory,

Kanchanbagh, Hyderabad-500258, India.

Abstract

A variety of composites based on MoSi$_2$ have been prepared by combustion synthesis or Self propagating High temperature Synthesis (SHS). A prior thermochemical analysis provides the adiabatic temperatures and the amounts of molten phases at the adiabatic temperature for the SHS of these composites. MoSi$_2$ - WSi$_2$ alloys, MoSi$_2$ - Mo$_5$Si$_3$, MoSi$_2$ - SiC, and (Mo,W)Si$_2$ - SiC composites were prepared by thermal explosion mode of combustion from elemental reactants. MoSi$_2$ - TiC composites were prepared by both thermal explosion and SHS modes of combustion. Phase formation has been confirmed by x-ray diffraction. In general the morphologies can be explained in terms of the calculated amounts of transient liquid phase present at the combustion temperature.

Introduction

Intermetallic composites in the systems Ni-Al (1,2), and Ti-Al (3) have been prepared by SHS reactions. Even though these aluminides have high specific strength, they are limited by low melting points and inferior oxidation resistance compared to silicides. Structural ceramics like SiC and Si$_3$N$_4$ are brittle from ambient temperature to elevated temperatures, resulting in unpredictable failure behaviour. Intermetallics based on silicides are emerging as competitors to both aluminides and structural ceramics, owing to their high melting points, excellent oxidation resistance and brittle-to-ductile transformation (BDT) (4-6). $MoSi_2$ becomes ductile above its BDT temperature of about 900°C. This property not only makes the failure behaviour of $MoSi_2$ more predictable but also makes it amenable to conventional metallurgical processing like hot deformation. However, due to the same property, above BDT, the strength of $MoSi_2$ decreases with increasing temperature. Hence both ambient temperature fracture toughness and elevated temperature strength need to be improved, before $MoSi_2$ can be put to application. Low temperature fracture toughness of $MoSi_2$ has been improved through ductile phase toughening (7-9). Elevated temperature strength can be improved by alloying with other silicides and/or reinforcing it with particulates or fibres (10-13).

Intermetallics can be prepared by elemental reactions using conventional furnace processing, SHS reactions or by the thermal explosion mode of SHS (14).

Conventional furnace processing

In the conventional furnace processing, the reactant powder compacts are heated at low heating rates to the required temperature and kept at these temperatures until the desired products are formed. At low heating rates, solid state diffusional reactions predominate, with low values of heat evolution. The reaction temperatures are well below the melting points of any of the reactants or products. Many times intermediate products form with incomplete conversion to the desired products. The products form a barrier between the reactants reducing the reaction rates considerably. Thus long reaction times are required for obtaining the desired products and the potential of the exothermic reactions is not realized.

SHS mode of combustion

In this mode of combustion, the reactant pellet is ignited at one corner, by a high energy heat input, and layer wise combustion occurs at a definite rate of wave propagation. The heat generated by the combustion of the first layer is conducted to the subsequent layers (precombustion zone) rapidly and hence the reaction rates are very high and the combustion is completed in a matter of few seconds, typical of SHS. Due to the very high heating rates, one or more of the reactants can be in the molten condition at the combustion temperature, resulting in such rapid reaction rates. The product can not form as a barrier between the reactants due to the presence of the liquid phase.

Thermal explosion mode of SHS

In this mode of combustion, a volumetric reaction occurs in the reactant compact when it is heated rapidly in a furnace. In the thermal explosion mode, the initial temperature can be so adjusted that the adiabatic temperature is above the melting point of not only some of the reactants but also some of the products. It was shown earlier that in the case of NiAl formation, if the product is also in the molten phase at the combustion temperature the velocity of propagation shows an abrupt increase by a factor more than three (15). Hence, such high reaction rates and complete conversion can be obtained by the thermal explosion mode of SHS.

Generally many intermetallics have low exothermic heat of reaction and as such supplementary heating is required. Hence these are prepared by thermal explosion mode of SHS. However $MoSi_2$ can be prepared either by SHS or thermal explosion mode of combustion. We have obtained $MoSi_2$ based alloys and composites by the following SHS reactions schemes (14,16-20):

$$Mo + 2\ Si \longrightarrow MoSi_2 \qquad \ldots (1)$$

$$(1+4x)\ Mo + (2+x)\ Si \longrightarrow (1-x).\ MoSi_2 + x.\ Mo_5Si_3 \qquad \ldots (2)$$

$$(1-x)\ Mo + x.\ W + 2\ Si \longrightarrow (1-x).\ MoSi_2 + x.\ WSi_2 \qquad \ldots (3)$$

$$(1-x).\ Mo + (2+x).\ Si + x.\ C \longrightarrow (1-x).\ MoSi_2 + x.\ SiC \qquad \ldots [4]$$

$$(1-x).\ Mo + 2(1-x).\ Si + x.\ Ti + x.\ C \longrightarrow (1-x).\ MoSi_2 + x.\ TiC \qquad \ldots [5]$$

By manipulation of the composition in the Mo-Si system, $MoSi_2 + Mo_5Si_3$ composites can be prepared according to Eq.2 (18). $MoSi_2$ - WSi_2 alloys (17), $MoSi_2$ - SiC (14,16), composites and a complex composite $(Mo,W)Si_2$ - SiC (19) were produced by thermal explosion mode of combustion. $MoSi_2$ - TiC composites were prepared by both thermal explosion and SHS mode of combustion (20). This paper presents a summary of results on the preparation of these composites.

Thermochemical Evaluation

In SHS reactions adiabatic temperature is an important parameter. It is the temperature to which the products are raised under adiabatic conditions as a consequence of evolution of heat due to chemical reaction. This temperature is a rough indicator whether a reaction is self propagating or not. Lower adiabatic temperatures can result in incomplete conversion or lead to oscillatory or spin combustion(21), while high adiabatic temperatures can result in rapid passing of combustion wave and hence incomplete conversion. Knowledge regarding the adiabatic temperatures and the amounts of molten phase formed at the adiabatic temperature helps in elucidating the reaction mechanism and product morphology. The adiabatic temperature can be calculated from the enthalpy of reaction, ΔH_{T_0} using the equation,

$$\Delta H_{T_0} = \int_{T_0}^{T_{ad}} C_p dT$$

where, T_0 is the initial temperature (the temperature at which the reaction is initiated), T_{ad} is the adiabatic temperature and C_p is the combined heat capacity of the products. The calculation procedure has been described earlier(22). From the calculation the amount of liquid phase formed can be obtained if the adiabatic temperature is above the melting point for any of the phases. The thermodynamic data required for the calculation is collected from standard reference books (23). Fig.1 shows variation of adiabatic temperature with increase in initial temperature for the formation of $MoSi_2$, WSi_2, Mo_5Si_3 and SiC. The initial temperature is the temperature to which the reactants are heated before ignition. The adiabatic temperature is arrested at the melting point of $MoSi_2$, WSi_2 and Mo_5Si_3. Adiabatic temperature for the formation of SiC continuously increases as no phase transformation is involved in the temperature range. For the formation of pure $MoSi_2$, the adiabatic temperature reaches the melting point of $MoSi_2$ for an initial temperature of 800 K resulting in partial melting of $MoSi_2$ and unless the initial temperature is raised above 1900 K the $MoSi_2$ formed is not fully molten. Similar values for the other phases can be obtained from Fig. 1. Fig.2 shows the variation of adiabatic temperature with the initial temperature for the formation of $MoSi_2$ (x=0), TiC (x=1) and various ($MoSi_2$ + TiC) composites (0 < x < 1) (20). For TiC formation, partial melting of TiC begins even when the reaction is initiated at room temperature and all the TiC formed is completely molten once the initial temperature exceeds 1400 K. For a composite of $MoSi_2$ with 0.6 mole fraction TiC, $MoSi_2$ is fully molten for an initial temperature of 400 K and TiC begins to melt only when the initial temperature is raised above 1650 K. Thus Fig. 2 gives the adiabatic temperatures for the formation of various composites of $MoSi_2$ - TiC.

Experimental

To prepare the $MoSi_2$ based composites by thermal explosion mode of combustion, required reactant powders were blended thoroughly, compacted and ignited by rapidly heating in an induction furnace, in inert atmosphere. For SHS mode of combustion, the compacts are ignited by a magnesium ribbon, under argon cover.

The phases formed were identified by x-ray diffraction using Mo Kα radiation. The morphology of the fracture surfaces of the products was examined using scanning electron microscope (JEOL JSM 840).

Results

Fig.3 shows the x-ray diffractograms, for $MoSi_2$, WSi_2 and their alloys prepared by combustion synthesis (17). It can be seen that only one set of x-ray peaks obtained for the alloys indicating continuous solid solution formation between $MoSi_2$ and WSi_2. This confirms the earlier reports on the solid solution formation between $MoSi_2$ and WSi_2 (12,24). Fig. 5 shows the x-ray diffractograms, for $MoSi_2$, SiC and their composites prepared by combustion synthesis (14). It can be seen that $MoSi_2$ and SiC show relevant peaks while the composites show mixed peaks and SiC peaks become prominent with increase in its content.

Fig.6 shows the diffractograms of the $(Mo,W)Si_2$ - 40 wt% SiC composite prepared from elemental powders by thermal explosion mode of combustion. Only one set of peaks form for $MoSi_2$ - WSi_2 solid solution and additional peaks for SiC are observed at all the three different heating rates used. Small peaks appeared for Mo_5Si_3 especially at higher heating rates. Mo_5Si_3 is supposed to improve the high temperature strength and low temperature toughness of $MoSi_2$ (18,25).

Fig. 7 shows the x-ray diffractograms of the $MoSi_2$ - TiC composites prepared from elemental powders by thermal explosion mode of combustion (20). The 40 wt% TiC composite shows clear peaks for $MoSi_2$ and TiC without any additional peaks, while the 20 wt% TiC composite show small additional peaks for Nowotny phase, $Mo_{<5}Si_3C_{<1}$. Fig. 8 shows the x-ray diffractograms of the $MoSi_2$ - TiC composites prepared by SHS mode of combustion. Here again the 40 wt% TiC composite shows negligible additional peaks, while the 20 wt% TiC composite show minor additional peaks for both Nowotny and ternary $(Ti,Mo)Si_2$ phases. No other binary phase is seen in any diffractogram as predicted by earlier thermodynamic calculation (20).

Fig.8 shows the morphology of $MoSi_2$, SiC and $MoSi_2$ - 20 wt% SiC composite. It can be seen that $MoSi_2$ shows a large particle morphology while SiC shows fine particle morphology. In case of composite, large agglomerated $MoSi_2$ can be seen along with fine SiC particulates. Fig. 9 shows the back scattered image of $MoSi_2$ - 20 wt% Mo_5Si_3 composite (18). Bright areas are Mo_5Si_3, grey areas $MoSi_2$ and dark areas are porosity. It is interesting to observe from Figs. 8 and 9, that the fracture morphology of $MoSi_2$ shows intergranular fracture while $MoSi_2$ - Mo_5Si_3 composite shows transgranular fracture mode. This is due to the presence of silica layer formed on the particles in case of $MoSi_2$ and the presence of eutectic phase prevents the formation of such silica layer in case of composite(18).

Fig. 10 shows the secondary and the corresponding backscattered electron images of the fractured surfaces of $MoSi_2$ - 40 wt% TiC composites prepared by thermal explosion and SHS modes of combustion (20). Backscattered electron images show an average atomic number contrast, and is very useful in analyzing these composites for uniform distribution of different phases. The light phase in these BSE images is $MoSi_2$, the grey phase is TiC and the dark phase is porosity. Here again it can be seen that in the thermal explosion mode $MoSi_2$ segregates and forms large particles while TiC is seen as fine particles. However the composite, prepared by SHS mode of combustion, shows a very homogeneous distribution of both phases as evident from the BSE image (20).

Discussion

The morphology of the combustion products depends upon the reaction mechanism. $MoSi_2$ based composites form generally by (solid + liquid) reaction, since at the combustion temperature, silicon is in liquid state while Mo is a solid phase. The experimental ignition temperature for preparing these composites is about 1573 K. As mentioned already, from Fig.1 it can be seen that except SiC, all other phases melt at the adiabatic temperatures corresponding to the initial temperature of 1573 K. This means that some amounts of transient molten phases are present at the combustion temperature for these phases and the morphologies presented in Figs. 8-10, reveal this fact. In case of $MoSi_2$ - SiC composites (Fig. 8), transient molten $MoSi_2$ particles segregate forming large $MoSi_2$ phase. In contrast, SiC forms fine particle morphology as there is no transformation involved for this phase (Fig.1). In the case of $MoSi_2$ - Mo_5Si_3 composites (Fig. 9), large particle morphology is formed due to the presence of transient eutectic phase at the combustion temperature.

In case of $MoSi_2$ - TiC composites prepared by thermal explosion mode of combustion (Fig. 10 a,b), the adiabatic temperature is 3154 K, which is the above the melting point of Mo and $MoSi_2$ but less than the melting point of TiC. Here the reaction mechanism is by (liquid +liquid) reaction and the adiabatic temperature is significantly higher than the melting point of $MoSi_2$. These may be some of the reasons for the severe segregation of $MoSi_2$ in this case. On the other hand for the $MoSi_2$ - TiC composites prepared by SHS mode of combustion, the adiabatic temperature is 2355 K, which is only slightly above the melting point of $MoSi_2$. In addition, diffusion times available in the SHS mode of combustion are always less than those for the thermal explosion mode, making segregation difficult. Due to these reasons $MoSi_2$ - TiC composite prepared by SHS mode of combustion shows highly homogeneous distribution (20).

Summary

$MoSi_2$ - WSi_2 alloys, simple composites of $MoSi_2$ with SiC, Mo_5Si_3, TiC and complex composite, $(Mo,W)Si_2$ - SiC are prepared from elemental powders by SHS. $MoSi_2$ - WSi_2 form continuous solid solutions. Simple composites with SiC and Mo_5Si_3 form desired product phases without any additional peaks. $MoSi_2$ - TiC and complex composite, $(Mo,W)Si_2$ - SiC show minor additional peaks. Transient molten phases present at combustion temperature generally contributes to the segregation in the thermal explosion mode of combustion. However, a highly homogeneous $MoSi_2$ - TiC composite has been obtained by SHS mode of combustion.

Acknowledgements

The authors are grateful to Director, Defence Metallurgical Research Laboratory, for his permission to publish this work.

References

1. S.D. Dunmead, Z.A. Munir, J.B. Holt and D.D. Kingman, J. Mat. Sci., 26, (1991)2410.

2. J.-P. Lebrat, A. Varma, and A.E. Miller, Metall. Trans., 23A,(1992)69.

3. L. Christodoulou, P.A. Parish and C.R. Crowe, in High temperature/High Performance Composites, ed. F.D. Lemkey, S.G. Fishman, A.G. Evans and J.R. Strife, (MRS, Pittsburg,) 120, (1988) 29.

4. A.K. Vasudevan and J.J. Petrovic, Mater. Sci. Engg., A155, (1992) 1.

5. J.J. Petrovic, MRS Bulletin, 7, (1993) 35.

6. K. Sadananda and C.R. Feng, Journal of Metals, (May 1993) 45.

7. E. Fitzer and W. Remmele, in Proc. Int. Conf. on Composite Mater. ICCM-V, ed. W.C. Harrigan,Jr., J. Strife and A.K. Dhingra AIME Publs.,Warrendale,Pa (1985), p 515 - 530

8. T.C. Lu, A.G. Evans, R.G. Hecht and R. Mehrabian, Acta Metall. Mater. 39, (1991)1853.

9. L. Xiao and R. Abbaschian, Metall. Trans. 23A,(1992) 2863.

10. J.M. Yang and S.M. Jeng, in Intermetallic matrix composites, ed. D.L. Anton, P.L. Martin, D.B. Miracle and R. McMeeking (MRS, Pittsburgh), 194, (1990) 139.

11. A.K. Bhattacharya and J.J. Petrovic, J. Am. Ceram. Soc. 75(1), (1992) 23 - 27.

12. J.J. Petrovic and R.E. Honnell, Ceram. Eng. Sci. Proc. 11 (7-8), (1990) 734.

13. J.J. Petrovic, A.K. Bhattacharya, R.E. Honnell, T.E. Mitchell, R.K. Wade and K.J. McClellan, Mater. Sci. Engg., A155, (1992) 259.

14. J. Subrahmanyam and R. Mohan Rao, J Am Ceram Soc., 78(2) (1995) 487.

15. V.M. Maslov, I.P. Borovinskaya and A.G. Merzhanov, Comb. Explos. Shock Wave USSR 12 (1976) 631.

16. J. Subrahmanyam, J. Am. Ceram. Soc. 76, (1993) C 226.

17. J. Subrahmanyam and R. Mohan Rao, Mater. Sci. Eng., A183, (1994) 203.

18. J. Subrahmanyam, J. Mater. Res., 9(10), (1994) 2620.

19. J. Subrahmanyam, R. Mohan Rao and K. Somaraju, Scripta Met. Mater., 31(10), (1994) 1317.

20. J. Subrahmanyam, R. Mohan Rao and G. Sundarasarma, J. Mater. Res., **10(5)**, (1995).

21. J. Subrahmanyam and M. Vijayakumar, Self propagating High-temperature synthesis - A review, J.Mater. Sci., 27, (1992) 6249.

22. J. Subrahmanyam, M. Vijyayakumar and S. Ranganath, Metals Materials and Processes, 1(2), (1989) 105.

23. J. Barin, O. Knacke and O. Kubaschewski, Thermophysical Properties of Inorganic Substances, Springer - Verlag (1973) and Supplement (1977).

24. J.O. Olowolafe, E.G. Colgan, C.J. Palmstrom and J.W. Mayer, Thin Solid Films, 138 (1986) 245.

25. R. Gibala, A.K. Ghosh, D.C. Van Aken, D.J. Srolvitz, A. Basu, H. Chang, D.P. mason, W. Yang, Mater. Sci. Eng., A155, (1992) 147.

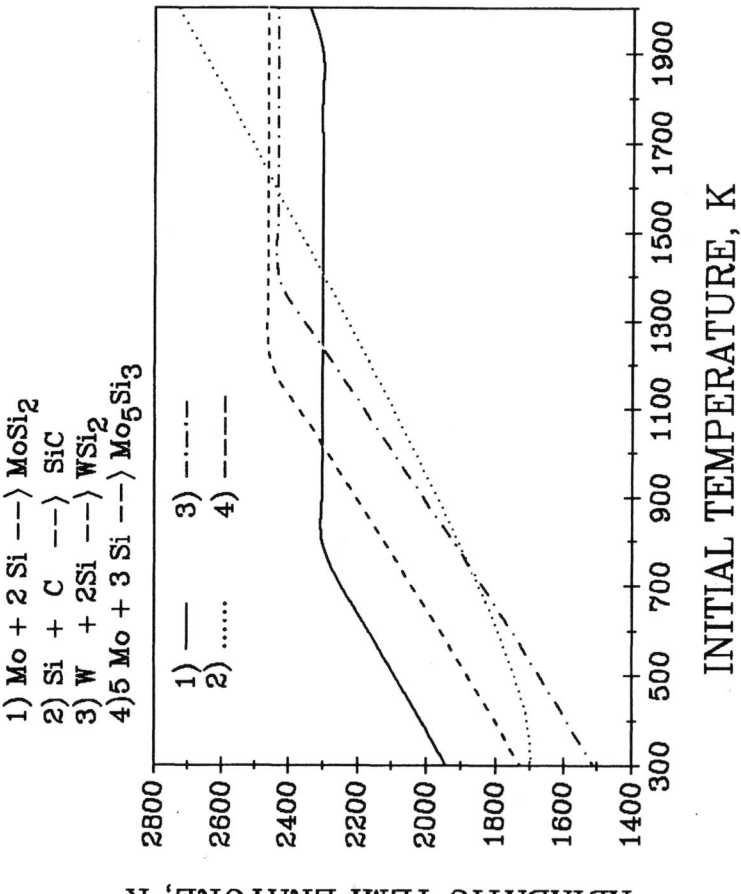

1. Calculated variation of the adiabatic temperature with initial reactant temperature for the formation of $MoSi_2$, WSi_2, Mo_5Si_3 and SiC.

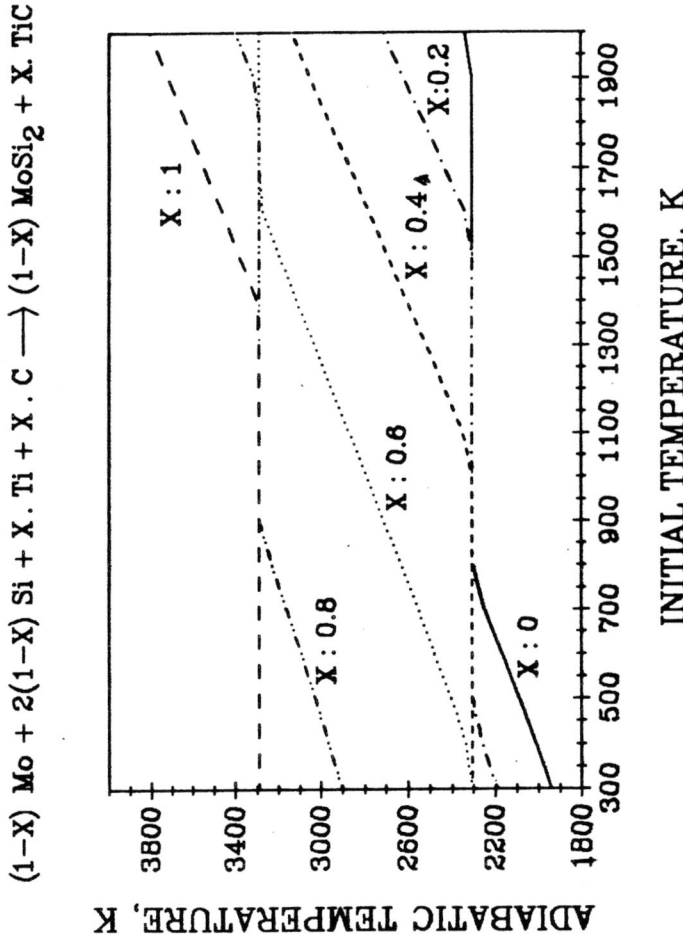

2. Calculated variation of the adiabatic temperature with initial reactant temperature for the formation of $MoSi_2$ - TiC composites.

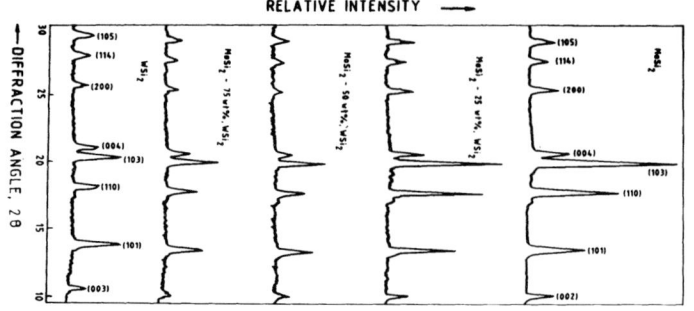

3. X-ray diffraction patterns for MoSi$_2$, WSi$_2$, and their alloys using Mo kα radiation, prepared by thermal explosion mode of combustion.

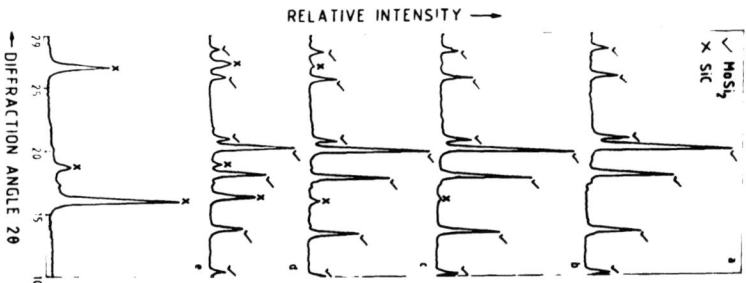

4. X-ray diffraction patterns for a) MoSi$_2$, for MoSi$_2$/SiC mole ratios of b) (3/1) c) (1/1), d) (1/4) and e) SiC (Mo Kα radiation), prepared by thermal explosion mode of combustion.

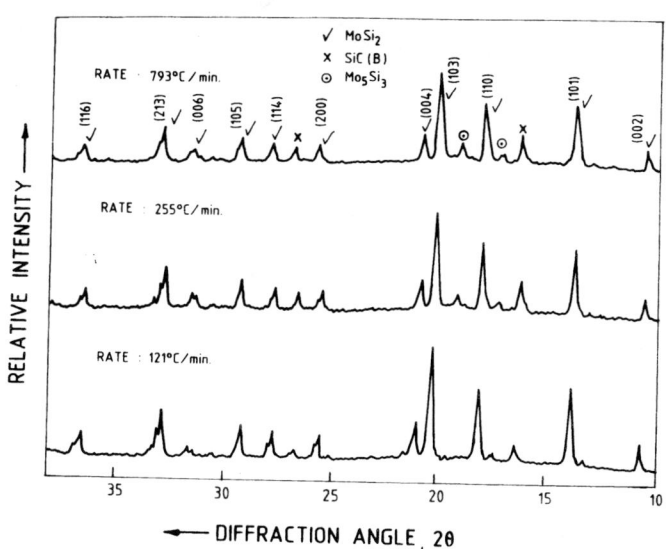

5. X-ray diffraction patterns for (Mo,W)Si$_2$ - 40 wt% SiC composite, prepared by thermal explosion mode of combustion, heated to combustion temperature at different heating rates.

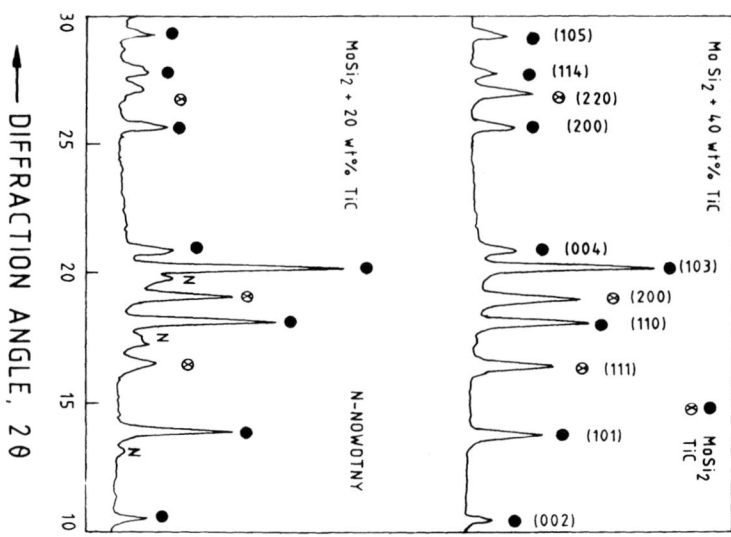

6. X-ray diffraction patterns for $MoSi_2$ - TiC composites prepared by thermal explosion mode of combustion from elemental powders.

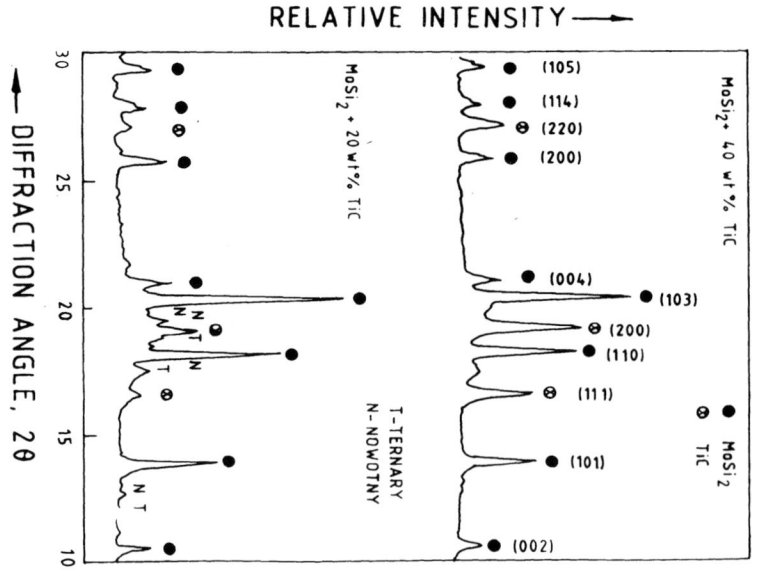

7. X-ray diffractograms of the $MoSi_2$ - TiC composites prepared by SHS mode of

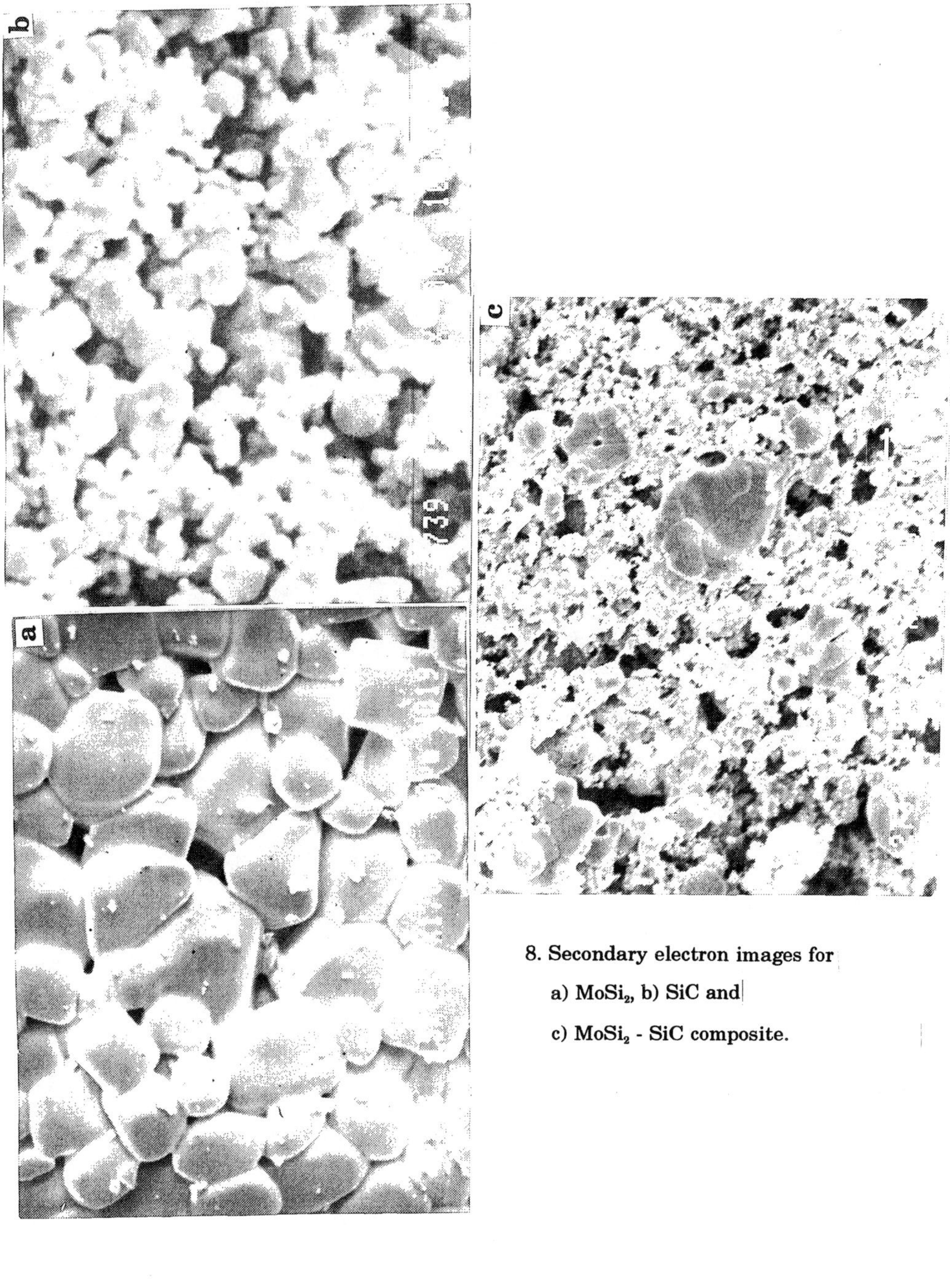

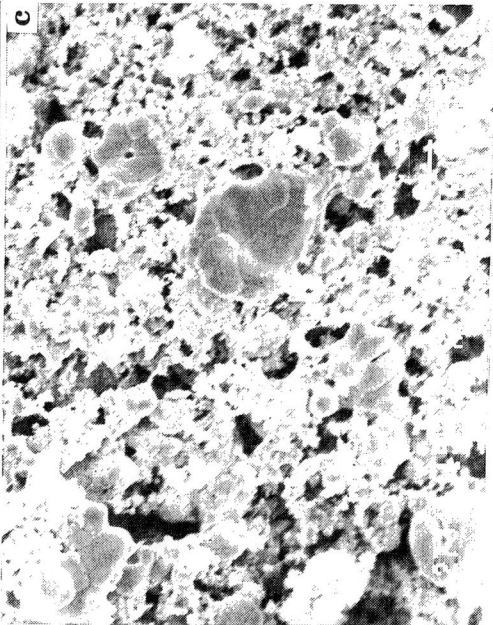

8. Secondary electron images for
a) $MoSi_2$, b) SiC and
c) $MoSi_2$ - SiC composite.

9. Back scattered electron image of the $MoSi_2$ - 20 wt% Mo_5Si_3 composite.

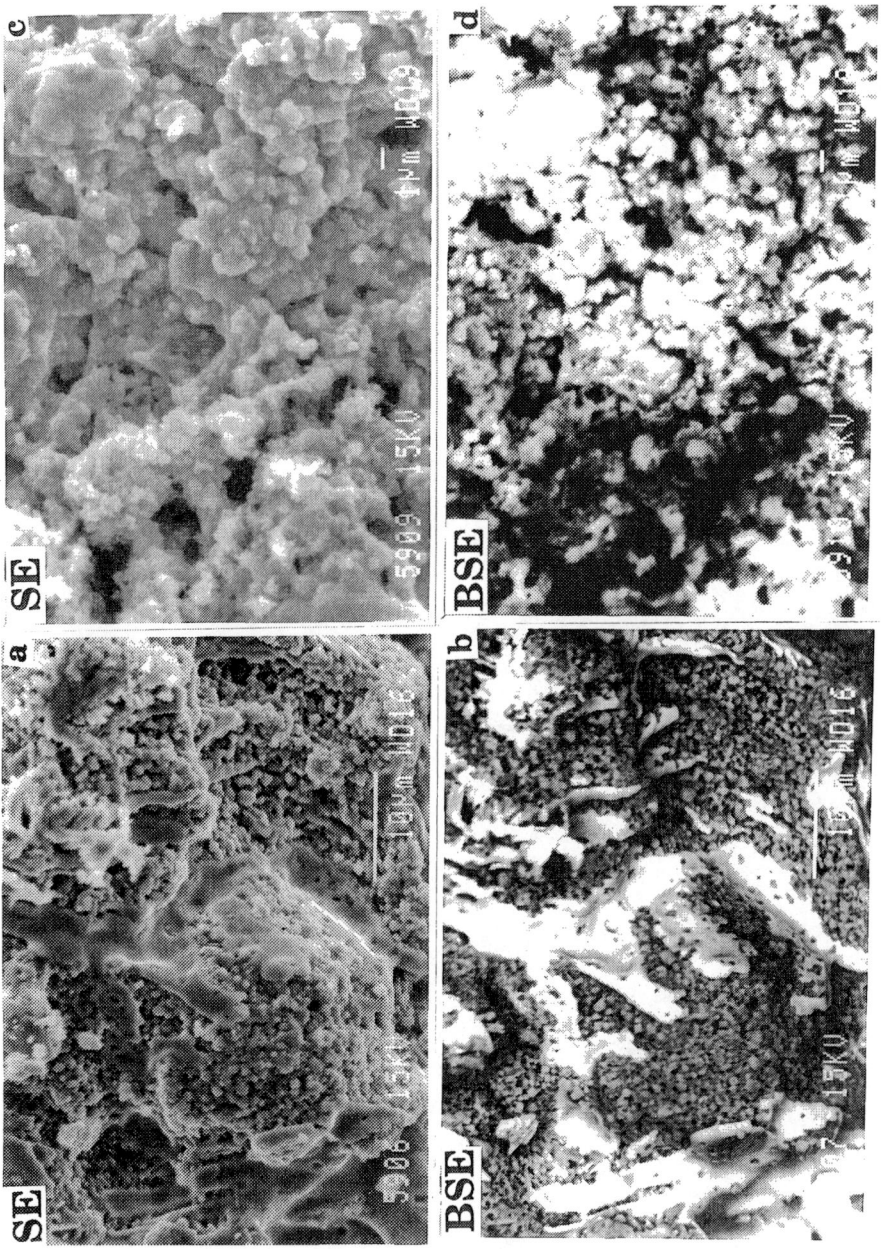

10. Secondary electron (SE) and the corresponding back scattered electron (BSE) images of the fractured surfaces of $MoSi_2$ - 40 wt% TiC composites prepared by thermal explosion mode of combustion (a,b) and SHS mode of combustion (c,d) from elemental powders.

DISCONTINUOUSLY REINFORCED TITANIUM MATRIX COMPOSITES via COMBUSTION-ASSISTED SYNTHESIS

S. Ranganath
Defence Metallurgical Research Laboratory, P.O. Kanchanbagh, Hyderabad 500058, India. Fax: 0091-040-239683

Abstract

Despite extensive research and development over the past two decades, titanium matrix composites have not realised their full commercial potential. For these composites the high value added costs of the available fabrication methods is largely due to the adverse liquid metal/reinforcement reactions. However, the DMRL combustion-assisted synthesis technique, which involves usual melting and casting of titanium in association with combustion synthesis reactions, has considerable potential for producing low cost composites with the required amounts of stable second phase reinforcements. This contribution discusses the novelty of this in situ technique to obtain a bimodal distribution (rod-like TiB and equiaxed Ti_2C) of reinforcements in the titanium matrix. Investigations reveal that reinforced composites possess ~30% higher stiffness than un-reinforced Ti. Tests also indicated that the composites were more deformable in compression and the values of yield strength increased with volume fraction of reinforcements, being approximately 3-5 times stronger than un-reinforced Ti. Additionally, it has also been shown that the measured steady state creep rate of the composite is 2-3 orders of magnitude lower than un-reinforced Ti in the investigated temperature range of 823-923 K.

I. INTRODUCTION

While the advantages of reinforcing aluminium with ceramic fibres and particulates are well known, the behaviour of discontinuously-reinforced titanium has not been extensively studied. From an engineering angle, the incorporation of low density, high modulus and high strength ceramic reinforcements can significantly enhance the specific modulus, specific strength and creep resistance of titanium metal and alloys[1-7]. Therefore, it is confidently predicted that titanium matrix composites will find a permanent place alongside conventional high temperature materials in the years to come. Current developments, such as missile fins, domed rocket cases and aircraft engine components, are strong pointers to future expectations.

If particulate-reinforced Ti-matrix composites are to gain industrial viability, there processing must be rendered economical and reliable. However, the high melting point of titanium (1933 K) and its high reactivity in the liquid state have turned out to be the major limitations in producing these composites[8-11]. The complexities are further accentuated because of the need to obtain a clean

Inorganic Matrix Composites
Edited by M. K. Surappa
The Minerals, Metals & Materials Society, 1996

matrix/particle interface which would have a beneficial effect on the overall mechanical properties.

In the past a few techniques such as cold and hot isostatic pressing[1] (CHIP), solidification processing[12-14] and exothermic dispersion[15,16](XD) routes have been utilised with measured success to produce Ti-matrix composites. But while the CHIP technique has some potential benefits in obtaining near-net shaped products, the capital investment is still higher compared to conventional and other emerging techniques. On the other hand, though the solidification and XD processing routes are inexpensive, there are some possible limitations with regard to chemical and thermal stability of a dispersed phase[17]. And both these routes involve the master alloy preparation and further remelting to prepare these composites, thus increasing the number of steps. Therefore, there is clearly a need to develop a single step process to obtain sound Ti-matrix composites. The newer processing route should not only be cost-effective but also flexible enough to produce composites with the required amounts of stable second phase reinforcements.

Keeping the aforementioned needs in mind, a novel technique[18] called combustion-assisted synthesis (CAS) has been developed in DMRL for Ti-based composites, which involves the usual melting and casting of titanium in association with combustion synthesis reaction. Using this technique it is possible to produce "designer microstructure" containing hard phases for strength, ductile phases for toughness and high aspect ratio reinforcements for creep resistance. The purpose of this paper is to explore the feasibility of synthesising titanium composites (in which titanium is reinforced with TiB + Ti$_2$C reinforcements) through CAS process and to evaluate the mechanical behaviour of reinforced and un-reinforced composites in the temperature range 298-873 K.

II. THERMOCHEMICAL EVALUATION

In combustion synthesis and combustion-assisted synthesis reactions, the adiabatic temperature is an important parameter. It is the temperature to which the products are raised under adiabatic conditions as a consequence of heat evolution due to the chemical reactions, and can be calculated from the enthalpy of the reaction, ΔH_{T_o} using the equation

$$\Delta H_{T_o} = \int_{T_o}^{T_{ad}} C_p \, dT \tag{1}$$

where T_o is the initial temperature at which the reaction is initiated, T_{ad} is the adiabatic temperature and C_p is the combined heat capacity of the products. The calculation of T_{ad} in the preparation of composites is described elsewhere[19,20].

One of the main problems encountered in combustion synthesis process, as applied to borides and carbides, is the process control. Sometimes the process heat is too high resulting in an explosion or it is too low to initiate the reaction. If the reaction heat is too high resulting in a very high T_{ad}, it can be reduced by the addition of an inert diluent, normally a constituent of the product itself. If T_{ad} is too low, supplementary heating is required. But when the T_{ad} calculated is higher than the melting point of any constituent of the product, then equation (1) is to be redefined to take this fact into consideration:

$$\Delta H_{T_o} = \int_{T_o}^{T_m} C_p \, dT + f \cdot \Delta H_m \qquad (2)$$

where ΔH_m is the enthalpy of fusion of the product having a melting point T_m, and f is the fraction molten. These principles are illustrated by performing thermodynamic calculations for the following reaction:

$$3\,Ti + B_4C = 2\,TiB_2 + TiC \quad \Delta H^o_{298} = -760.2 \text{ kJ} \qquad (3)$$

where ΔH^o_{298} is the standard heat of reaction at 298 K.

Fig. 1 shows the effect of initial temperature on the adiabatic temperature and the fraction melted. About 45% of TiB_2 phase is in molten condition when the sample is ignited at room temperature. If all the TiB_2 formed is to be in the molten condition, the reactant temperature is to be raised to 900 K. TiC phase formed in the reaction starts melting only when the initial temperature is raised to about 1100 K and complete melting of TiC phase occurs when the initial temperature is above 1400 K. It should be noted that if the ignition occurs below these temperatures, it is no longer possible to control the reaction.

The preparation of titanium matrix composites through CAS process requires the presence of additional moles of titanium. Once this excess titanium is available, the governing reaction for the preparation of Ti-TiB-Ti$_2$C composites is

$$(x + 6)\,Ti + B_4C = x\,Ti + 4\,TiB + TiC \qquad (4)$$

Fig. 2 shows the variation of adiabatic temperature with initial temperature for 10, 20 and 30 excess moles of titanium for the above reaction. It can be seen that the adiabatic temperature reaches the melting point of Ti for 10, 20 and 30 excess moles of Ti at the initial temperatures of 800 K, 1200 K and 1350 K respectively. But the excess moles of Ti will be fully molten only at initial temperatures of 1100 K, 1550 K and 1800 K. Depending on these excess moles of Ti, the respective initial temperatures have to be reached to keep the additional titanium in molten condition. This would ensure uniform distribution of TiB + Ti$_2$C reinforcements. However, if the adiabatic temperature exceeds 2500 K the decomposition of TiB occurs. Hence the excess moles of Ti have to be closely monitored.

III. EXPERIMENTAL

For preparing a control sample of titanium, titanium sponge was used. In a Leybold Haraeus make melting set up, non-consumable vacuum arc melting was carried out to obtain pancakes of control sample of Ti.

For processing TiB + Ti$_2$C reinforced composites, the raw materials used were titanium sponge, titanium and boron carbide powders. For composite preparation, Ti and B$_4$C powder mixture was used in the form of cylindrical compacts compacted to 60% theoretical density. To get 10, 15 and 20 vol.% of second phase reinforcements in titanium matrix, titanium sponge and the required amount of compacts were together initiated according to reaction (4) through a special technique[18]. Thus the exothermic heat evolved as a consequence of ignition of the compact is effectively utilised in melting the excess titanium, resulting in composites with dispersoid of TiB and Ti$_2$C.

The control sample and the composites were hot rolled (50% reduction at 1073 K) and annealed (1073 K). For determining the Young's modulus of control sample and the composites, the measurements of the ultrasonic velocities through the specimens (10 mm height and 10 mm diameter) were performed using the pulse-echo overlap method. To study the composite behaviour at room temperature under tensile loading, samples of 4 mm gauge diameter and 25 mm gauge length were machined from the flat rolled plates with the specimen axis parallel to the rolling direction. Two specimens were tested for each composite in the Instron machine, and load-displacement plots obtained were converted to true stress-true plastic strain curves. Yield strength values were taken for 0.2% off-set strain. For the compression tests, samples of 10 mm diameter were cut. Tests were conducted in the temperature range 298-873 K. A split-type resistance heating tubular furnace was used for high temperature tests and temperature was maintained within ±1 K.

A constant-load creep machine with compression grip was used to study the steady state creep behaviour of the composites. Compression specimens were machined in the form of solid cylinders (5 mm gauge diameter and 10 mm gauge length) from the flat rolled plates with the specimen axis parallel to the rolling direction. The creep strain of the sample was measured using two parallel linear variable displacement transducers (LVDTs) mounted on the ridges provided on the compression grips. The temperature of the sample was monitored using three separate thermocouple tied, one at the specimen and the other two at the grips. The specimen temperature was controlled within ±1 K.

Auger electron spectroscopy (AES) and scanning electron microscopy was performed on deep etched specimens, as well as on fracture surfaces. Transmission electron microscopy (TEM) was performed at 300 kV on specimens thinned by ion milling.

IV. RESULTS AND DISCUSSION

A. *Microstructure*

The AES back scattered electron image of 20 vol.% composite is shown in Fig. 3. It reveals that the second phase reinforcements are either rod-like or lower aspect ratio/equiaxed particles. It is seen that the rod-like particles are rich in boron and equiaxed ones are rich in carbon; further, electron probe microanalysis on the reinforcements confirmed them to be TiB and Ti_2C respectively.

A typical deep-etched microstructure of the composite is shown in Fig. 4. The high aspect ratio reinforcements are now clearly visible in this figure. It was also observed that in this route of processing the morphology of the reinforcements remained same in 10, 15 and 20 vol.% composites[20].

In CAS the second phase reinforcements are formed in situ and, therefore, the general problems encountered in powder metallurgical approaches and conventional molten metal techniques are totally averted. But the presence of reinforcements can drastically reduce the grain size of the matrix by pinning the grain boundary movement. This aspect is clearly seen in Fig. 5, where the matrix grain size is reduced to 1-2 μm in composites, compared to ~170 μm of un-reinforced titanium[20]. This in situ process also results in a clean matrix/particle interface (Fig. 6). The reduction in grain size and a clean matrix/particle interface is bound to influence the stiffness, strength and creep behaviour of the composites.

B. *Mechanical Properties*

The Young's moduli of the control sample and the composites are given in Table I. It is to be noted that the moduli of the composites is 20-30% higher than the un-reinforced material and increases with increasing reinforcement content.

The room-temperature tensile and compressive stress-strain curves obtained for 15 vol.% composite is shown in Fig. 7. It is seen that the compressive yield strength of the composite is higher than its tensile yield strength. From these curves it is obvious that the total deformation level for the composites is extended to more than 10% in compression compared with <3% in tension, and that the composites are stronger in compression at all strains. Back scattered electron micrograph of the composite taken in the vicinity of fractured area from the cross-section of tensile specimen is shown in Fig. 8. It is observed that fracture occurs due to particle cracking in preference to particle decohesion from the matrix. This confirms that in the composites produced by CAS route there is a strong interfacial bonding between the particle and the matrix because of a clean interface.

In Fig. 9 yield stress of composites is plotted against test temperature at a constant strain rate. The yield stress is seen to decrease with increasing temperature. The significant aspect is that the composite strength is significantly higher than the un-reinforced titanium even at 873 K. In fact, the ratio of composite strength to un-reinforced material strength increases from 2-3 at 298 K to 3.5-5 at 873 K. This indicates that the potential benefit in strengthening due to reinforcements is greater at higher temperatures.

C. *Steady State Creep Behaviour*

The variation of steady state creep rate with stress for the control sample and the composite are plotted in Fig. 10(a) and 10(b), both on logarithmic scales. The data can be fitted by straight lines with slopes equal to the value of the stress exponent,n. This indicates that the data follows power law equation

$$\dot{\varepsilon} = \frac{A E b D_0}{kT} \left(\frac{\sigma}{E}\right)^n \exp\left(-\frac{Q}{RT}\right) \qquad (5)$$

where $\dot{\varepsilon}$ is the creep rate, σ is the applied stress, E is Young's modulus, Q is activation energy, R is gas constant, k is Boltzmann constant, T is temperature, D_o is frequency factor for diffusion, b is Burgers vector and A is a dimensionless constant. For un-reinforced Ti and 10 vol.% composite n is more or less same (between 4 - 5), yet the creep rate of composite is found to be 2-3 orders of magnitude lower than Ti in the investigated temperature range. Also, for a steady state creep rate of 1×10^{-7} s^{-1}, it is noticed in Fig. 11 that increasing the volume fraction of reinforcements increases the creep strength of titanium composite.

As seen from a representative TEM micrograph of 15 vol.% composite (Fig. 12), the interface between the particle and the matrix is clean even after creep deformation at 873 K. The grain size of the matrix in the as-crept sample was also observed to be unchanged[20,21]. It is important to note that the stress exponent is similar for the matrix and the composites, but the composites have higher creep strength. This strengthening can originate from (a) load transfer to stiffer phase, and/or (b) microstructure strengthening. Load transfer to the stiffer phase results in an overall increase in modulus. From equation (5) it is clear that increase in modulus would lower the creep rate of the material.

In un-reinforced titanium with large grains, the dislocation creep is governed by the generation and mutual annihilation of dislocations within the grain. But if the grains are reduced to such an extent (as in Ti-composites) that there could be no intragranular sources present, then the boundaries would become the source and the sinks for dislocations and a microstructural dependence might arise. In Fig. 13, which shows the bright field and weak beam dark field micrograph of as-crept composite, the movement of dislocation across the grain and absence of any dislocation-dislocation interaction is evident. This gives conclusive proof of the microstructural strengthening concept.

V. CONCLUSIONS

Based on the above studies, the following conclusions are presented.

(1) A new technique for the processing and production of TiB + Ti_2C reinforced composites have been developed.

(2) The thermodynamic calculations of adiabatic temperature with excess moles of titanium helps in defining the appropriate conditions for the preparation of composites by this technique. To obtain sound Ti-TiB-Ti_2C composites, it is necessary that the adiabatic temperature should be >1933 K and initial temperature be less than the decomposition temperature of TiB.

(3) Compared with the control samples, Young's modulus of 10-20 vol.% composites increased by 20-30% and the yield strength increased by 100-200%. These property enhancements have been attributed primarily to the reduction in grain size and a clean matrix/particle interface.

(4) Composite strength has been shown to be significantly higher than the un-reinforced titanium in the investigated temperature range of 298-873 K.

(5) Compressive creep tests reveal that the creep rate of composites is 2-3 orders of magnitude lower than un-reinforced Ti in the temperature range 823-923 K.

(6) Both control sample and the composites show power-law creep behaviour with stress exponents between 4 - 5. In this case the strengthening of composites has been attributed to its higher modulus and microstructural strengthening.

In summary, the present study was oriented towards examining the effects of TiB + Ti_2C reinforcements on a ductile Ti matrix. Though the development of Ti-matrix composites is still in its infancy stage and the evolutionary process for them has only just begun, it would be worthwhile examining the effects of the reinforcements on a high strength alloy-matrix like Ti-6Al-4V or TiAl. This is likely to pay good dividends in terms of high-temperature applications.

ACKNOWLEDGMENTS

Program support from the Defence Research and Development Organisation of India and the permission of Director, DMRL to publish this paper are gratefully acknowledged.

REFERENCES

1. S. Abkowitz and P. Weihrauch: *Adv. Mat. Proc.*, 1989, vol. 136(1), pp. 31-34.
2. M.H. Loretto and D.G. Konitzer: *Metall. Trans.*, 1990, vol. 21A, pp. 1579-87.
3. T.P. Johnson, J.W. Brooks and M.H. Loretto: *Scripta Metall. Mater.*, 1991, vol. 25, pp. 785-89.
4. S.J. Zhu, Y.X. Lu, Z.G. Wang and J. Bi: *J. Mat. Sci. Lett.*, 1992, vol. 11, pp. 630-32.
5. S. Ranganath, M. Vijayakumar and J. Subrahmanyam: *Mat. Sci. Engg.*, 1992, vol. A149, pp. 253-57.
6. S. Ranganath, T. Roy and R.S. Mishra: DMRL, India, unpublished research, 1994.
7. S. Ranganath and S.V. Kamat: DMRL, India, unpublished research, 1995.
8. P. Soumeldis, J.M. Quenisset, R. Naslain and N.S. Stoloff: *J. Mat. Sci.*, 1986, vol. 21, 895-903.
9. D.G. Konitzer and M.H. Loretto: *Acta Metall.*, 1989, vol. 37, pp. 397-406.
10. S.K. Choi, M. Chandrasekharan and M.J. Brabers: *J. Mat. Sci.*, 1990, vol. 25, 1957-64.
11. J.C. Rawers, W.R. Wrzesinski, E.K. Roub and R.R. Brown: *Mat. Sci. Tech.*, 1990, vol. 6, pp. 187-91.
12. J. Chen, Z. Geng and B.A. Chin: in *High Temperature Ordered Intermetallic Alloys III*, MRS Symp. Proc., MRS, Pittsburgh, PA, 1989, vol. 133, pp. 447-52.
13. R. Zee, C. Yang, Y. Lin and B. Chin: *J. Mat. Sci.*, 1991, vol. 26, pp. 3853-61.
14. M.E. Hyman, C. McCullough, J.J. Valencia, C.G. Levi and R. Mehrabian: *Metall. Trans.*, 1989, vol. 20A, 1847-59.
15. A.R.C. Westwood: *Metall Trans.*, 1988, vol. 19A, 749-58.
16. L. Christodoulou, P.A. Parrish and C.R. Crowe: in *High Temperature/High Performance Composites*, MRS Symp. Proc., MRS, Pittsburgh, PA, 1988, vol. 120, pp. 29-34.

17. D. Lewis: in *Metal Matrix Composites: Processing and Interfaces*, R.K. Everett and R. J. Arsenault, eds., Academic press, New York, NY, 1991, pp. 121-32.

18. S. Ranganath: Process for Producing Particulate-Reinforced Titanium Matrix Composites, Patent Filed, India, August 1994.

19. J. Subrahmanyam, M. Vijayakumar and S. Ranganath: *Met. Mater. Proc.*, 1989, vol. 1(2), 105-11.

20. S. Ranganath: *Combustion-Assisted Synthesis and Evaluation of Titanium Matrix Composites*, Ph.D. Dissertation, Banaras Hindu University (India), 1993.

21. S. Ranganath and R.S. Mishra: DMRL, India, unpublished research, 1994.

Table I Measured Moduli of Ti-TiB-Ti$_x$C Composites

Vol.% Reinforcements	Modulus (GPa)
0	110
10	129
15	136
20	145

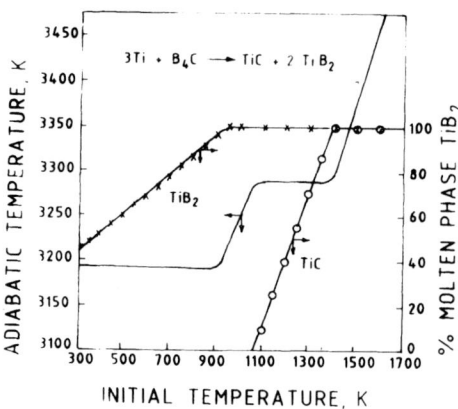

Fig. 1: Effect of initial temperature on the adiabatic temperature and per cent molten phase (TiB$_2$) for the reaction 3 Ti + B$_4$C = TiC + 2 TiB$_2$.

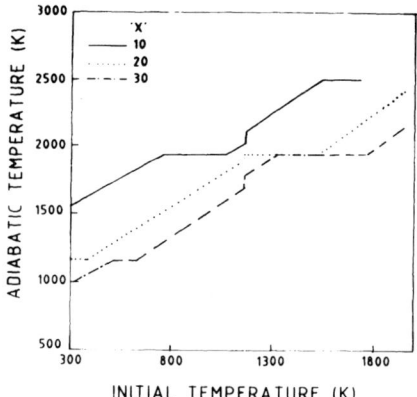

Fig. 2: The dependence of adiabatic temperature on initial temperature for the reaction (x + 6) Ti + B$_4$C = x Ti + 4 TiB + Ti$_2$C.

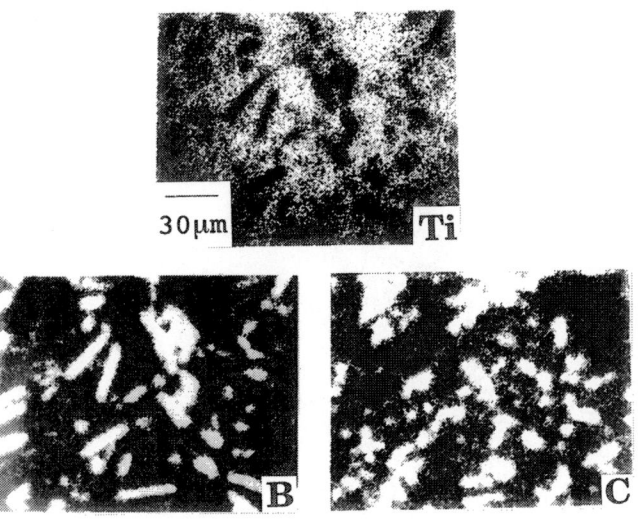

Fig. 3: Auger elemental maps of Ti-20 vol.% TiB + Ti$_2$C composite.

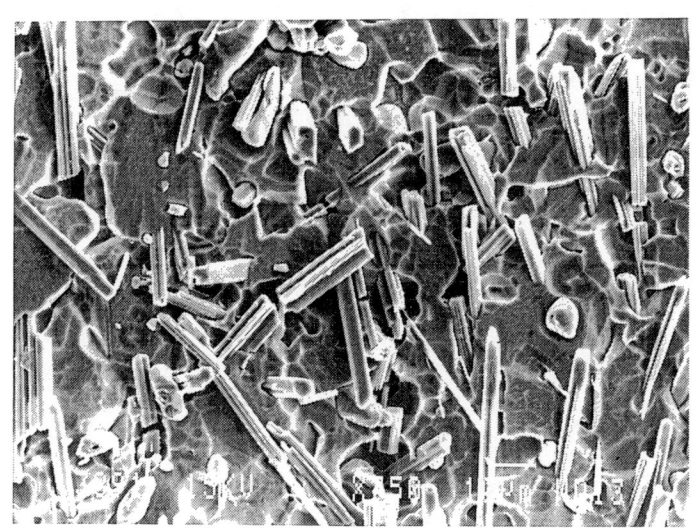

Fig. 4: Deep etched SEM micrograph of 15 vol.% composite exhibiting the presence of high aspect ratio TiB reinforcements.

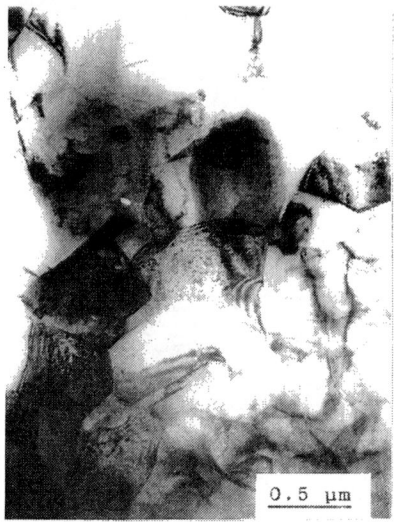

Fig. 5: Bright field TEM micrograph of 15 vol.% composite showing the matrix grains.

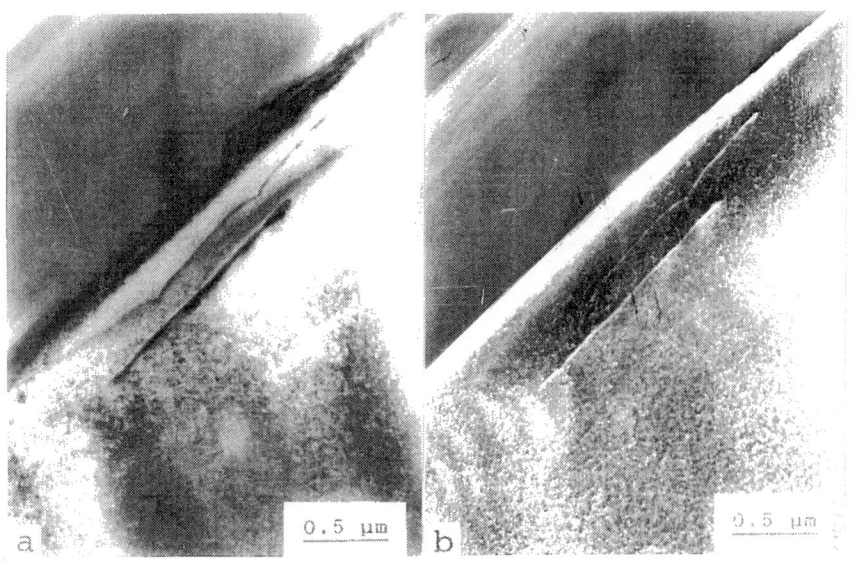

Fig. 6: Clean matrix/particle interface in 15 vol.% composite: (a) bright field, and (b) dark field.

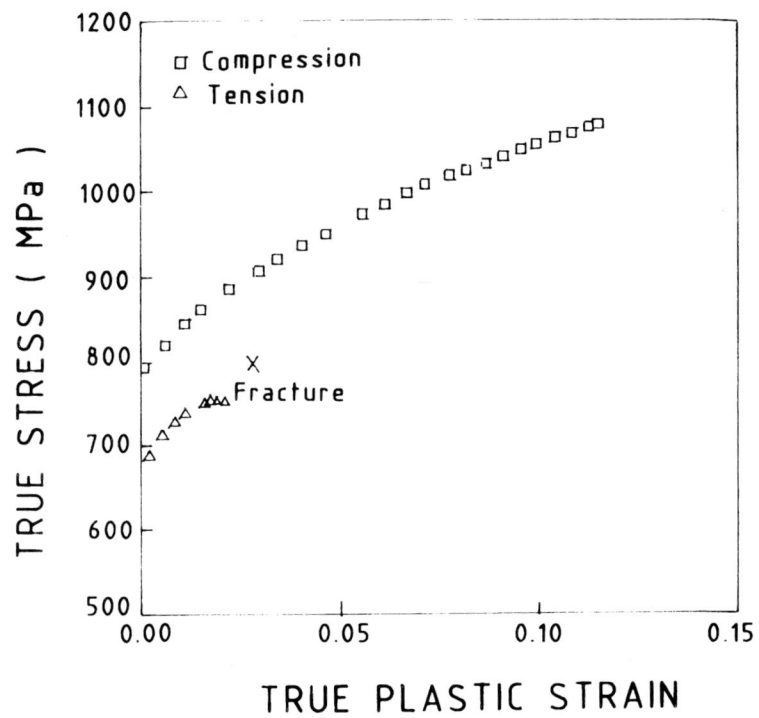

Fig. 7: Room temperature tensile and compressive true stress-true plastic strain curves for 15 vol.% composite at a strain rate of 2×10^{-3} s^{-1}.

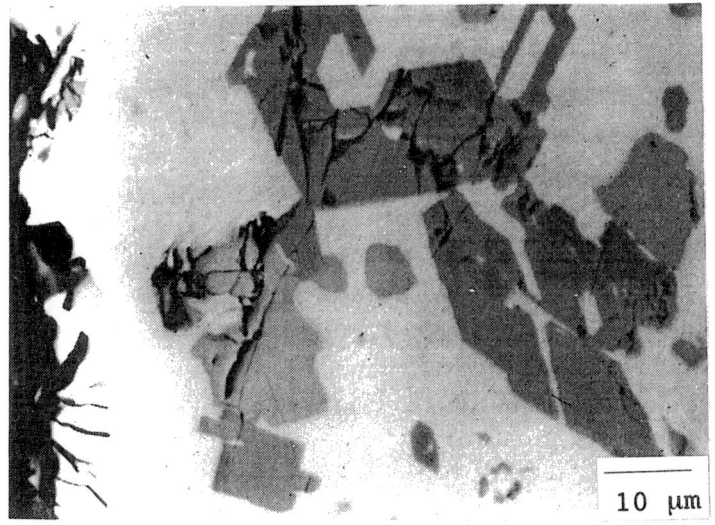

Fig. 8: Scanning electron micrograph taken using back scattered electrons on a cross-section of fracture surface of 20 vol.% composite.

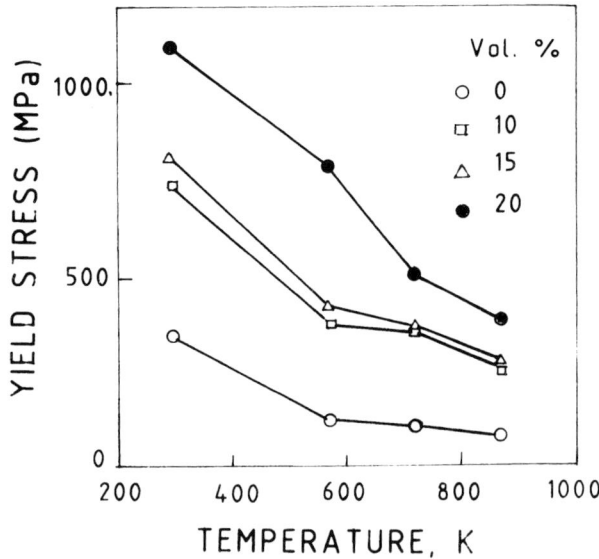

Fig. 9: Yield stress vs. temperature plots for 0-20 vol.% composites at a strain rate of 2×10^{-3} s^{-1}.

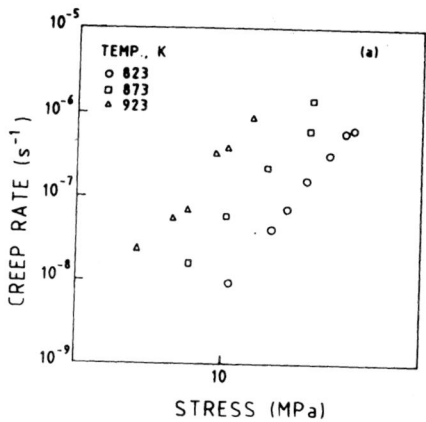

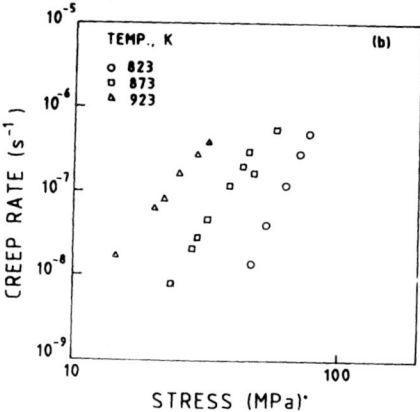

Fig. 10: The variation of steady state creep rate as a function of applied stress for (a) control sample, and (b) 10 vol.% TiB + Ti$_2$C composite.

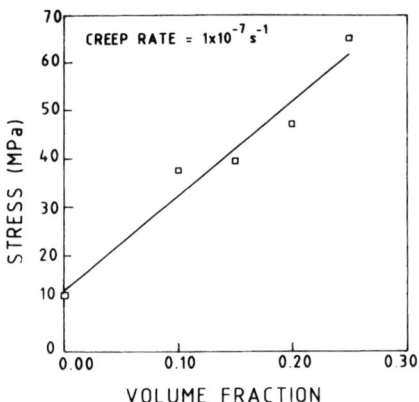

Fig. 11: The variation of stress as a function of volume fraction of reinforcement.

Fig. 12: Bright field TEM image showing a clean matrix/particle interface in an as-crept 15 vol.% composite.

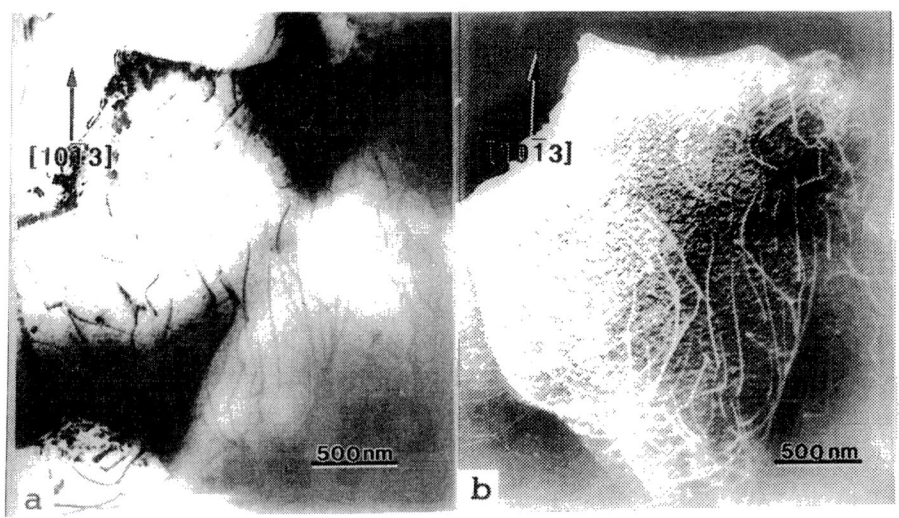

Fig. 13: (a) Bright field and (b) dark field TEM images from one grain of 10 vol.% TiB + Ti_2C composite specimen, observed after creep at 873 K and 46.6 MPa.

"Production and Characterization of MMC tubes"

Anil Kumar Gupta, Rajiv Sikand, R.C. Anandani, Ajay Dhar and I.A. Malik

Metals & Alloys Group
National Physical Laboratory
Dr. K.S. Krishnan Marg
New Delhi - 110012

ABSTRACT

Hot extrusion process has been used to convert aluminium alloy-SiCp MMC billets into rods and tubes for specific end applications. Billets were produced by both liquid metallurgy route and powder metallurgy techniques. In the former, the use of canning the billet, prior to extrusion, is necessary to avoid serious surface defects such as porosity, gas entrapments, etc, in the extrudants. In terms of properties, the hot extruded products from PM processed billets yielded much better mechanical properties as compared to the ones using LM processed billets. In Powder Metallurgy technique, with the use of hot pressing the billets, prior to extrusion, the properties are enhanced, especially ductility, and this process is cost effective. Products have been characterized to determine the property levels achievable in the rods and tubes which show promise. The various aspects pertaining to deformation process in Discontinuously Reinforced Aluminium MMCs (DRA) are also reviewed and some application of DRA tubes are presented.

I. INTRODUCTION

The attractive physical and mechanical properties that can be obtained with Metal Matrix Composites (MMCs), such as higher specific modulus, strength, thermal stability and improved tribological properties, over their conventional monolithic counterparts, enables the material scientists to find its applications in automobile, aerospace and general engineering industries. MMCs combine metallic properties (ductility & toughness) with ceramic properties (high thermal stability & high modulus) leading to greater strength in shear & compression and higher service temperature capabilities.

Discontinuously reinforced MMCs have, therefore, recently attracted considerable attention as a result of the following:

(a) availability of various types of reinforcements at competitive costs

(b) successful development of manufacturing processes including Powder Metallurgy (PM), Spray Atomization & Deposition (SAAD) and Liquid Metallurgy (LM) to produce MMCs with reproducible structures and properties

(c) offer essentially isotropic properties with substantial improvements in strength and stiffness

(d) availability of standard or near standard metal forming processes (forging, extrusion, rolling & superplastic forming) which can be utilized to form these MMCs into usable products.

Numerous techniques have been developed for the production of discontinuously reinforced MMCs which can be broadly classified as:

(a) powder metallurgy, i.e., solid phase process

(b) liquid metallurgy, i.e., liquid phase process, and

(c) spray atomization and deposition technique, i.e., two

phase (solid/liquid) process.

Each process has its own merits and demerits and thus finds applications in one area or the other. The selection of the process mainly depends upon the performance requirement, matrix-reinforcement compatibility and costs.

Once this material is produced in an ingot form, the problem remains of converting this material into usable shapes and sizes with improved properties. This can be achieved by employing metal forming techniques. There are different processes which can be used but the important ones are hot extrusion, hot rolling, hot forging and closed die forging.

The stiffness, strength and fracture characteristics of MMCs have a direct influence on the interface between the matrix and the reinforcement. A 'good' interface means a strong chemical bond across the interface, which efficiently transfers the stresses from the ductile matrix to the strong and stiffer reinforcement. The increment in the Young's modulus achieved by incorporating ceramic reinforcement in a metallic matrix, is a measure of the interfacial bond integrity between the matrix and reinforcement. A well bonded interface would allow the maximum transfer of load from the matrix to the reinforcement through the interface and is expected to exhibit a high value of E_{expt}/E_{rom} ratio, where E_{expt} and E_{rom} are the Young's modulus obtained experimentally and from the rule of mixtures, respectively. The property-performance characteristics of MMCs are critically dependent on the interface bonding between the matrix and the reinforcement and these secondary processing techniques have a direct influence on this property.

In the present paper various aspects pertaining to the deformation process in Discontinuously Reinforced Aluminium MMCs are presented. In particular, the details of hot extrusion process employed to produce Al-alloy SiCp composite rods and tubes, from billets synthesized from both PM and LM processes, are discussed in detail. The matrix and reinforcement used are Al alloy (2124 & 6061) and SiCp, respectively. Some application of these MMC tubes are also discussed briefly.

II HOT DEFORMATION BEHAVIOUR

Hot workability of DRA is one of the most important research topic currently being pursued by various research groups all over the world. Hot working process above the recrystallization temperature not only deforms the material in required shapes and sizes but also refines the microstructure which in turn improves the mechanical properties such as ductility and strength. Another important feature which influences the hot workability is the nature of distribution of the ceramic reinforcement in the matrix material. A non-uniform distribution (clustering of reinforcement) may lead to inhomogeneous flow of the material leading to cracks while a uniform distribution of ceramic reinforcement and equiaxed matrix grain structure leads to homogeneous flow of material. As a result, the load requirements are higher in the former.

However, the process parameters adopted for the standard matrix (monolithic) alloys can not be directly employed for processing the respective composite materials, due to the incompatible deformation characteristics of the matrix and the reinforcement particulates, as well as the consequent changes in the optimal processing parameters.

The incompatibility between a deforming matrix and a nondeformable particle means that dislocations will be generated at the particles. Humphreys et al.[1] have studied the microstructure developed and recrystallization behaviour of 1050 Al reinforced with 2-35 vol% of SiCp in the size range of 1-100 μm. In addition to dislocation generation due to deformation incompatibility, the mismatch in the coefficients of thermal expansion of the matrix and the reinforcement also gives rise to dislocation generation. In addition to the high dislocation densities, the presence of particulates (> 1 μm) also leads to particle stimulated nucleation (PSN) of recrystallization in the deformation zones adjacent to the particles [2,3]. As the ceramic particles used in composites are generally much larger than 1μm in size, these will be a function of the particle size and volume fraction. A high volume fraction of fine particles with close interparticle spacing can also pin the grain boundary migration and thereby stabilize the fine grain structure and restrict grain growth. Especially in powder metallurgy composites, the oxide particles exert an additional effect on the kinetics of the recrystallization and grain growth[1].

DRA composites produced by either powder metallurgy or liquid metallurgy routes, need to be deformed to achieve the best possible properties. In the case of the powder metallurgy billets, the prior particle boundary (PPB) defects can be eliminated and the distribution of the reinforcement particles are improved by the hot deformation process. The matrix powder particles, especially aluminium powder consist of an oxide layer, which does not allow proper interface bonding between matrix and reinforcement as well as between matrix particles themselves. These can only be broken and redistributed by hot deformation using extrusion & forging involving extensive shear. In the case of powder metallurgy billets, the reinforcement particles usually being finer, as compared to the matrix particles, are distributed around the matrix powder particles forming a necklace like microstructure in the as-hot pressed composites. Although, this situation can be improved to some extent by choosing much closer size-ratio of matrix and reinforcement particles, significant improvement can only be achieved by a hot deformation step involving considerable matrix material flow around the hard reinforcement particles, thereby redistributing them and obtaining a more uniform distribution. Even in the case of composites produced by the liquid metallurgy technique, deformation processing helps in healing the defects like porosities and segregation to a large extent. It also breaks the cast dendritic structure and replaces it with a fine grained microstructure.

A. Processing maps in hot working

Deformation processing of MMCs presents its own problems, due to the incompatible deformation characteristics of the matrix and reinforcements. Hot working of MMCs thus requires to be carried out within a certain 'window' of strain rate and temperature in order to achieve defect free parts, microstructure and mechanical properties.

Optimization of hot working processing parameters by trial and error is both time consuming and expensive. Different methods such as hot torsion testing, hot tensile testing and actual deformation processing studies can also be used to understand the deformation characteristics. Xia et al.[4,5] have adopted hot torsion testing and extensive transmission electron microscopy to study the hot deformation and recrystallization behaviour of 6061 Al/SiCp composites. They have concluded that the higher flow stresses encountered in composites, as compared to the monolithic alloys, is due to the higher dislocation densities due to CTE and the denser substructure induced by plastic flow around the rigid SiC particle during hot working. The softening effects at high temperature deformation is attributed to increased dynamic recovery and dynamic recrystallization.

Processing maps developed by Prasad et al.[6] provide a scientific tool for optimizing process parameters based on simple laboratory hot compression tests. Processing maps are generated based on the dynamic materials model (DMM) developed by Prasad et al.[6] and reviewed by Gegel et al.[7]. The model considers the workpiece as a dissipator of power and the power input to the workpiece is partitioned and dissipated as viscoplastic heat and microstructural changes. The strain rate sensitivity (SRS) 'm' is found to be responsible for the partitioning of power and a dimensionless parameter 'n' is defined as the efficiency of power dissipation through microstructural processes and derived to be equal to 2m/m+1. This efficiency parameter is plotted as a function of both strain rate and temperature to get the processing map. The different efficiency peaks in the map are then interpreted in terms of different deformation mechanisms and thus help in identifying the safe domains for processing.

Bhat et al.[8] have studied the hot working characteristics of DRA composites in detail. The conventional Al alloys having good formability are pressed at high strain rates but their respective composites exhibit flow instabilities at those strain rates. Flow instabilities manifest themselves as extensive cracking at particle/matrix interfaces, adiabatic shear band forming and localized flow due to adiabatic heating.

III. EXPERIMENTAL DETAILS

Several R & D organizations & institutes in India are engaged in the development of MMCs, employing LM and PM routes. The National Physical Laboratory (NPL), New Delhi is collaborating with Defence Metallurgical Research Laboratory (DMRL), Hyderabad for PM route, Hindustan Aeronautics Limited (HAL), Bangalore, Regional Research Laboratory (RRL), Trivandrum, for LM route. Hot deformation behaviour of these materials, using different techniques are being studied at NPL, New Delhi where the facilities exist to study different characteristics using hot extrusion and closed die forging on a 500-ton vertical hydraulic press, which is shown in Fig.1.

A. Powder Metallurgy

Conventional powder metallurgy technique involving series of steps is shown in Fig.2.

This process involves blending of SiCp (average particle sizes of 1.9, 3.0 & 14 μm) with 2124 Aluminium alloy powder (average particle size 94 μm) in varying volume fractions (10, 15 & 20 %). These blended powders were degassed and subsequently Cold Isostatically Pressed (CIP). These CIPed compacts were then vacuum hot pressed at a temperature above the solidus of the alloy to achieve 98 ~ 99% theoretical density. Alternatively we have also adopted an "modified" PM process for making billets for extrusion by simply hot pressing the "green" MMC compacts, thereby eliminating vacuum hot pressing, which is the most expensive process in the conventional powder metallurgy technique for making billets.

These billets, obtained from "conventional" and "modified" PM process, were subsequently hot extruded. The mechanical and metallurgical properties of different components are reported in the later section of the paper.

B. Liquid Metallurgy

The matrix material and reinforcements used in the present investigations were 2124/6061 Al-alloy and SiCp (10, 15 & 20%) particulates of average size 23 & 40 μm. DRA billets containing various volume fractions of SiCp were prepared using LM route (stir casting). This technique essentially involves addition of the dispersoid particles into the vortex formed in a pool of mechanically stirred molten alloy at 730 ~ 740°C. SiCp particulates were sieved and preheated in a separate furnace for about 3 hours at 775°C before addition to the melt. After the reinforcement was incorporated into the melt, stirring was continued for 15 minutes and then cast into permanent moulds. The stirring time, after the incorporation of SiCp, was kept minimum to prevent any reaction between SiCp and liquid metal. Different composite billets were produced using varying volume fractions (10 to 20%).

IV. EXPERIMENTAL RESULTS

The MMC billets employing PM and LM techniques were hot extruded forged at NPL, New Delhi, using different extrusion conditions. Different process parameters such as temperature, pressing speed, extrusion ratio, die design, lubrication and its effect on varying volume fraction of various compositions of DRA have been optimized mainly to investigate their effect on surface quality, pressure requirement and the resultant mechanical/ metallurgical properties. Details of experimentation techniques used are reported elsewhere[9,10]. The work was extended to produce circular rods, tubes and shapes. These circular DRA tubes find applications in automobile and space, and have been discussed in later section.

A. PM Route

72 mm diameter PM processed billets were hot extruded using a 500 ton press. Different extrusion parameters like extrusion ratio, billet temperature, die design, lubrication, etc., were optimized to produce satisfactory extruded rods, tubes and shapes. These products were characterized for detailed mechanical and metallurgical properties. Fig.3 shows the pressure encountered for carrying out solid extrusions employing different process parameters. Fig.4 shows the pictorial view of these DRA circular rods with different extrusion ratios (9:1, 16:1 & 36:1).

These extruded products exhibit excellent surface finish and are free from any surface defects or porosity. Microstructure showing well distributed SiCp reinforcement in the matrix is shown in Fig.5. Tensile tests were conducted on these samples using an Instron tensile testing machine. Comparison of properties achieved using different extrusion conditions are shown in Table 1. All the properties reported are in as-extruded conditions and are likely to be considerably improved in T6 condition. Billets produced by the "modified" PM route were also extruded and resulted in the enhancement of mechanical properties, especially ductility, as reported in Table 2.

This work has been undertaken with an objective to make circular tubes as structural component for space and automobile applications. Typical properties achieved for these DRA tubes are shown in Table 3, and Fig.6 shows the pictorial view of these tubes.

B. LM Route

Extrusions were made from the DRA billets (6061 Al + SiCp and HMS 2112 Al + chopped carbon fibres) to produce MMC rods and tubes. Various extrusion process parameters were optimized to obtain good surface finish and good metallurgical and mechanical properties of these extruded products. In addition to these materials, Cu-graphite[11], Al-flyash, Lead-flyash, have also been secondary processed to circular rods at NPL.

These tubes show considerable surface defects, especially in tubes with higher volume fraction of reinforcement. The metallographic examination suggests fairly uniform distribution of SiCp and some of the particles showed orientation in the direction of extrusion. The particle crowding was also observed with the increase of SiCp content in the matrix. Table 4 shows the mechanical properties of these extruded tubes for different SiCp content. It is quite clear from this table that the mechanical properties of these extruded LM processed billets are quite inferior, especially ductility, as compared to powder metallurgy. This is due to the presence of defects like gas entrapments in the extruded tubes which are not healed even after extrusion.

All the rods and tubes extruded using PM processed billets gave better properties as compared to ones obtained from the LM processed billets. Furthermore, the mechanical properties can further be improved by using "modified" PM processed billets for extrusion. Efforts are underway to optimize different processing conditions to further improve the properties. However, one of the main problems in hot extrusion of MMCs is the high wear of extrusion die/container. The problem is more acute while processing LM route billets because of coarser reinforcements used. Another major problem, which still needs to be addressed, is the high rate of wear of the tool bits/extrusion saw blades. Efforts are being made to improve their performances.

V. APPLICATIONS OF DRA TUBES, DEVELOPED AT NPL, FOR SPACE & AUTOMOTIVES

While the United States is taking lead in the use of MMCs in defence and aerospace, Japan has already introduced this new material for a large number of applications in automotives. Other countries which have strong activities in this area are, U.K., France, Germany, Netherlands, Australia, Canada and China[12]. Number of Indian R & D institutes and industries are actively engaged in the development of this hi-tech material, studying different aspects of the science & technology of these composites. In spite of these developments, this material has still to find its place in the Indian industries. Some of the opportunities and challenges in this context are also highlighted in later sections.

A. MMC tubes for satellite applications

DRA are now emerging as cost effective materials, providing designers with a unique combination of design features and properties such as, high specific stiffness and strength, controlled coefficient of thermal expansion (CTE), improved elevated temperature properties and superior dimensional stability. Tubular struts for satellites is one such application where high specific stiffness, low CTE and reasonably high thermal conductivity play an important role. The performance indices for optimal selection of materials for the tubular strut application are both, resistant to mechanical ($E^{1/2}/\rho$), as well as thermal (k/CTE) distortions, where, ρ is the density, E the elastic modulus, k the thermal conductivity, and CTE is the coefficient of thermal expansion. MMCs exhibit significantly higher figures of merit as compared to the corresponding unreinforced alloy as shown in Table 5. The percent weight savings which could be achieved by substituting conventional 2124-Al alloy with Al-Li alloy or MMCs are indicated in Table 6.

Table 6 indicates that the use of MMC tubular struts can offer significant weight savings which is a prime requirement for the satellite applications. The properties achieved for the tubes developed at NPL, are mentioned in Table 3. Efforts have been made to improve the ductility of these tubes to approximately 5%.

B. MMC tubes as drive shafts

Drive shafts in passenger cars offer particularly attractive applications for DRA. The primary driving force for aluminium drive shafts is its significant lower weight coupled with the ease with which it can be balanced. The critical drive shaft speed (Nc) is given as[13]:

$$Nc = 15\pi/L^2 \ [(E/\rho)g(R_0 + R_1)^2]^{1/2}$$

where, L is drive shaft length, R_0 the outer radius and R_1 the inner radius, E the elastic modulus, ρ the density and g is the gravitational acceleration. As is clear the specific modulus (E/ρ) is the only material property that effects the critical speed. Some future vehicles will require drive shafts that are longer than feasible using aluminium or steel. Both steel and aluminium have low specific modulus and the critical speed limitations of these imposes severe constraints on the drive shaft lengths. DRA drive shafts represent a cost effective solution to this problem as the specific modulus of DRA can be significantly higher than that of aluminium or steel.

Manufacturing technologies for the fabrication of drive shafts include consistent material quality, high tolerance tube extrusions and DRA to aluminium fusion welding techniques. Significant utilization of these drive shafts are expected by the late 1990s.

C. DRA tubes as cylinder liners

The main emphasis in automobile industries is to reduce vehicle weight. A typical 'rule of thumb' estimates that 10% reduction in vehicle weight yields approximately 5.5% improvements in fuel economy[11]. To achieve this, conventional cast iron engine blocks are being replaced with aluminium silicon alloy engine blocks. A few of the companies have adopted an interim approach of using cast iron blocks fitted with DRA liners. This will help in achieving improved thermal stability, reduce engine friction by improving block stiffness and dimensional stability. These liners increase weight reduction and wear resistance. The extruded DRA tubes at NPL have been fitted on trial basis in cast iron engine blocks (Fig.7) for a two-wheeler and the performance trials are in progress. These DRA circular tubes also find applications for bicycle frames and golf clubs.

VI. OPPORTUNITIES AND CHALLENGES - INDIAN CONTEXT

Despite the fact that MMCs are emerging as potential materials for different applications in automobile and general engineering industries, a number of barriers have to be overcome to ensure its widespread introduction. A few of these could be classified as below:

* MMC finished components will eventually cost more than conventional components because of the higher cost associated with raw materials and processing costs (fabrication and machining).
* recycling of this material is another handicap and needs attention.
* lack of design data, particularly the detailed properties of fatigue, creep, wear and corrosion is again a practical problem and requires a comprehensive data generation.
* lack of quantitative information on machining tool life and die wear.
* product consistency is also an important parameter. Standards are required so that automotive engineers could specify and manufacture components with specified structure and properties that are predictable and stable overlong periods.

To overcome these deficiencies, sustained efforts are required to develop near-net shaped forming techniques, rapid and inexpensive machining processes and recycling methods.

ACKNOWLEDGEMENTS

The author is thankful to Professor E.S.R. Gopal, Director, NPL for his permission to publish this work. Special thanks are also due to Dr. Y.R. Mahajan, DMRL, Hyderabad and Professor E.S. Dwarakadasa of IISc, Bangalore with whom we have a very successful and fruitful collaboration.

REFERENCES

1. F.J. Humphreys, W.S. Miller and M.R. Djazeb: *Mat. Sci. & Tech.*, 1990, vol. 6, p. 1157.
2. F.J. Humphreys: *Acta. Metall.*, 1977, vol. 25, p. 1323.
3. F.J. Humphreys: *Proc. Conf. Recrystallization '90'*, (ed. T. Chandra), Warrendale, PA, TMS, (1990), p. 113.
4. X. Xia, P. Sakaris and H.J. McQueen: *Proc. of 'Composite Materials (ICCM-9)*, Vol. 1, (ed. A. Miravete), Woodhead Publishing Ltd, Spain, (1993)
5. X. Xia, P. Sakaris and H.J. McQueen: *Mater. Sci. Technol.*, 1994, vol. 10, p. 487
6. Y.V.R.K. Prasad, H.L. Gegel, S.M. Doraivelu, J.C. Malas and J.T. Morgan: *Metall. Trans. A*, 1984, vol. 15A, p. 1883.
7. H.L. Gegel, J.C. Malas, S.M. Doraivelu and V.A. Shende: *Metals Handbook*, Vol. 14, ASM International, Materials Park, OH, (1987) p. 417
8. B. V. Radhakrishna Bhat, Y. R. Mahajan, H. Md. Roshan and Y.V.R.K. Prasad: *Metall. Trans. A*, 1992, vol. 23A, p. 2223.
9. A.K. Gupta, K.E. Hughes, and C.M. Sellars: *Met. Technol.*, 1980, vol. 3, p. 323.
10. A.K. Gupta and K.E. Hughes: *Metallurgia*, 1979, vol. 46, p. 644.
11. M. Gupta, R. Sikand and A.K. Gupta: *Scripta Metall. Mater.*, 1994, vol. 30, p 1343.
12. Pradeep K. Rohatgi: *Defence Science Journal*, 1993, vol. 43, p. 323.
13. John E. Allison and Gerald S. Cole: *J.O.M.*, January (1993), p. 19.

Table 1

Tensile results of as-extruded rods, using PM processed billets.

[2124 Al + 20% SiCp (14 μm)]

ER	Temp. (°C)	YS (MPa)	UTS (MPa)	%Elong.	E (GPa)	Hardness (VHN)
9:1	450	190	270	3.0	90.7	91.6
	500	197	280	3.3	92.8	
16:1	450	195	275	3.1	93.3	94.2
	500	181	291	3.7	96.0	
36:1	450	210	310	3.0	95.9	105
	500	217	316	3.7	98.8	

2124 Al + 20% SiCp (1.9 μm)

ER	Temp. (°C)	YS (MPa)	UTS (MPa)	%Elong.	E (GPa)
36:1	525	300	397	4.8	100

Table 2

Tensile properties of as-extruded rods, using PM processed billets

[2124 Al + 20% SiCp (14 μm)]

ER	Temp. (°C)	YS (MPa)	UTS (MPa)	%Elong.	Modulus* (GPa)
36:1	525	198	323	7.9	95.56
36:1	500	226	333	5.6	95.65

* Mean of eight readings, obtained by Elastosonic equipment

Table 3

Tensile properties of extruded tubes (T6), using PM processed billets.

[2124 Al/15% SiCp (1.9 μm)]

Property	Result
YS (MPa)	380
UTS (MPa)	473
% Elongation	3.6
E (GPa)	102

Buckling properties (L/D ratio: 2.5)

Property	Result
Compressive YS	261 (MPa)
Ultimate Buckling stress	453 (MPa)

Table 4

Tensile properties of extruded tubes, using LM processed billets

[2124 Al/SiCp (23 μm)]

Sample	U.T.S (MPa)	%Elong.
6061 Al/10 wt% SiCp	266	0-2
6061 Al/15 wt% SiCp	207	-
6061 Al/20 wt% SiCp	260	-

Table 5

Figures of merit for the selection of material for tubular strut applications

Material	E (GPa)	ρ (g/cm^3)	$E^{1/2}/\rho$ (Mech.)	k/CTE (Thermal)
2124-Al	72	2.78	3.05	5.96
2124Al + 15%SiCp	100	2.84	3.52	7.38
2124Al + 30%SiCp	130	2.86	3.98	8.80

Table 6

Weight savings by use of MMCs and Al-Li alloys for structural applications (failure mode: Buckling)

Material (Baseline - 2124 Al)	% Weight Savings $[\rho_2/\rho_1(E_1/E_2)^{1/2}]$
8090 Al-Li	13.7
2124 Al/15%SiCp	17.6
2124 Al/30%SiCp	19.6

Fig.1 :   A pictorial view of the 500-ton vertical hydraulic multipurpose press

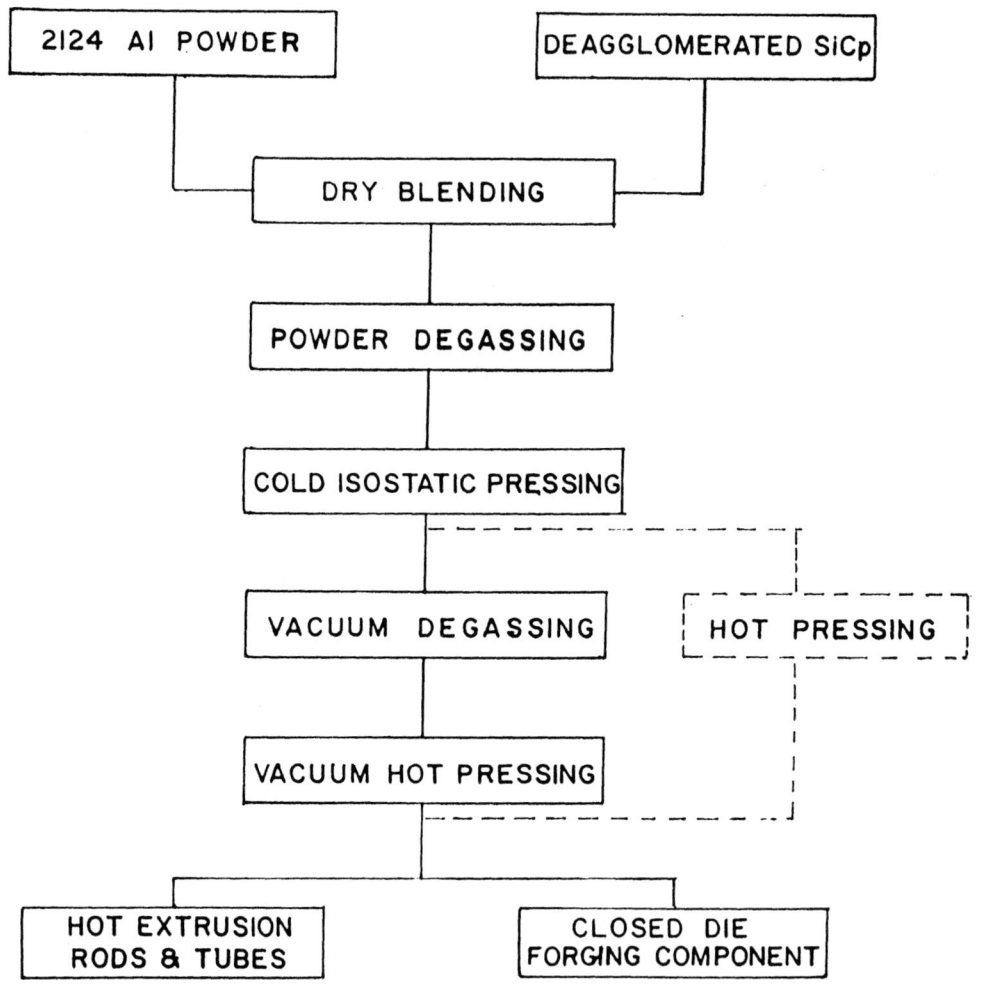

Fig.2 : PM process flow chart for production of billets. The dotted line indicates the "modified process" for making these billets for extrusion.

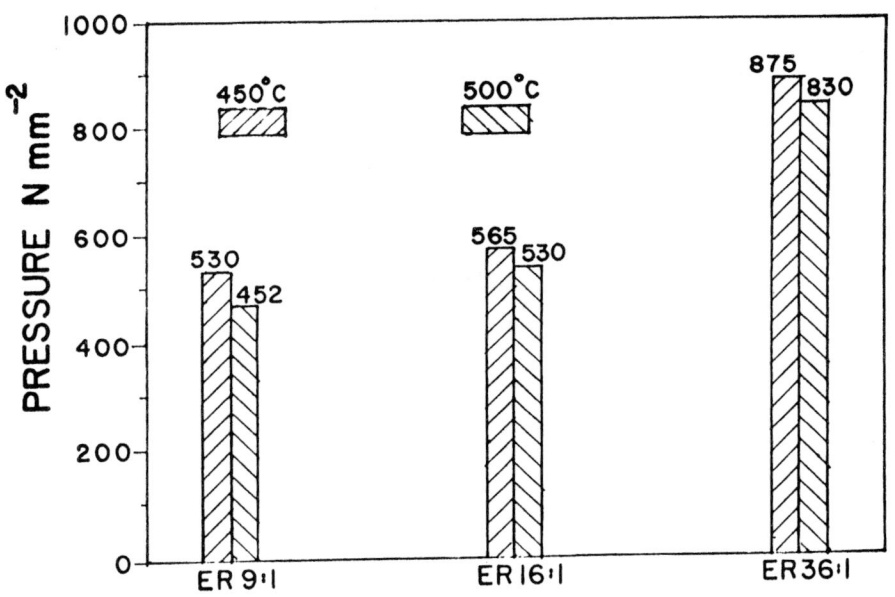

Fig.3 : Effect of Extrusion ratio and temperature on pressure requirement for extrusion.

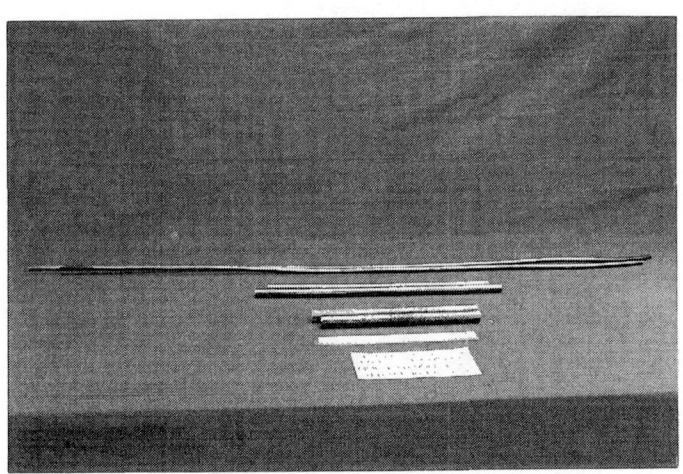

Fig.4 : Extruded DRA rods using different extrusion ratios.

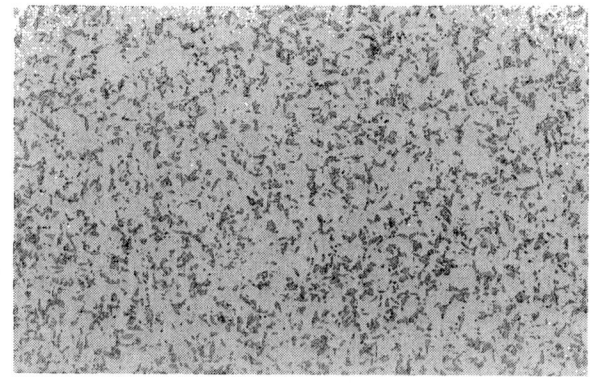

Fig.5 : Typical micrograph of an extruded rod (X100).

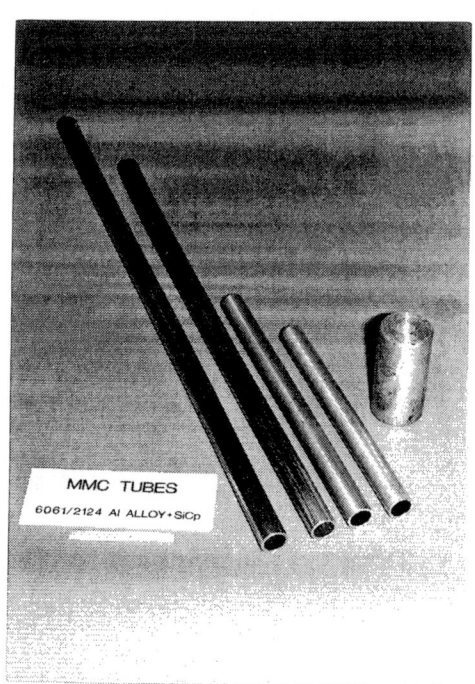

Fig.6 : Extruded DRA circular tubes.

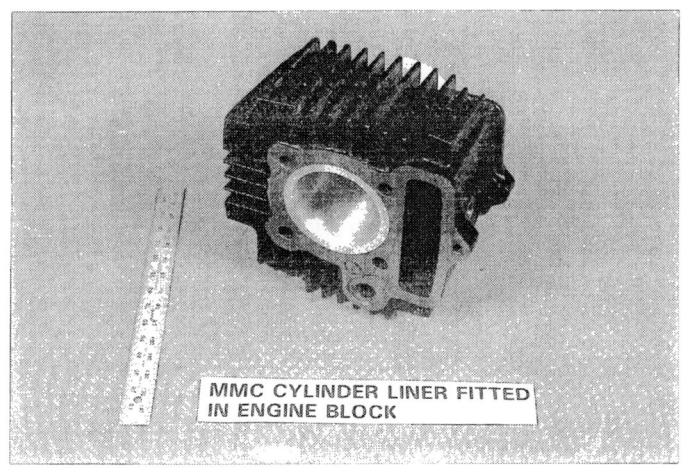

Fig.7 :   DRA cylinder liner in an engine block.

DISCONTINUOUSLY REINFORCED PM PROCESSED MMCs ACTIVITY AT THE DEFENCE METALLURGICAL RESEARCH LABORATORY, HYDERABAD-500 058, INDIA

B.V.R.BHAT, V.V.BHANUPRASAD, R.MITRA, M.K.JAIN, A.B.PANDEY AND Y.R.MAHAJAN

Abstract

The Defence Metallurgical Research Laboratory, Hyderabad, India is pursuing an extensive research program aimed at developing the powder metallurgy approach of producing discontinuously reinforced metal matrix composites. Apart from optimizing the different processing steps, the MMCs have been characterized for mechanical & physical properties and microstructure. Al alloy matrix composites reinforced with silicon carbide particulate have been developed for applications requiring dimensional stability as one of the important design parameter. Deformation processing of these composites have been studied and the optimal parameters identified by constructing processing maps based on the dynamic materials model. Processing maps have also been used in understanding the deformation mechanisms and the effect of matrix material, reinforcement volume fraction and the initial microstructure on the same. The role of interfaces was investigated and different reinforcements and processing routes studied to improve the interface characteristics. The creep behaviour of these MMCs has also been evaluated and a dislocation creep mechanism map proposed. The effect of different reinforcements, particle sizes and volume fractions on the creep resistance of MMCs has been studied. $MoSi_2$ matrix composite reinforced with SiC powder has been developed by the reaction hot pressing route and characterized. Some of the important results in all these areas are discussed in this paper.

Inorganic Matrix Composites
Edited by M. K. Surappa
The Minerals, Metals & Materials Society, 1996

I. INTRODUCTION

Discontinuously reinforced aluminium (DRA) composites produced by the powder metallurgy route are the focus of an extensive research and development program undertaken at the Defence Metallurgical Research Laboratory (DMRL), Hyderabad, India. As the name implies, DRA composites are aluminium based materials reinforced by particulate, whisker or short fibres. The reasons for their success are their desirable properties, including low density, high specific stiffness, controlled coefficient of thermal expansion, superior dimensional stability and increased fatigue resistance. The isotropic properties offered by these composites, their amenability to conventional metal working processes unlike continuous fibre reinforced composites, easy availability of cheaper reinforcements, along with the availability of comparatively low-cost, high volume production methods are the additional advantages.

The activity at DMRL includes both basic research and component development programs. The powder metallurgy process involving different steps like deagglomeration, blending, degassing, cold isostatic pressing, hot pressing, deformation processing etc., has been first optimized on a small scale on pure aluminium matrix composites[1,2] and later on using 2124 Al alloy matrix[3]. The processing steps have been scaled up by establishing facilities for each step, the most important one being the setting up of a 200 Tonne Vacuum Hot Press. Dimensionally stable MMCs have been developed for applications such as laser mirror substrate and characterized for important design parameters like micro yield strength. Deformation processing of these MMCs is important for achieving better distribution of the second phase particles, elimination of prior particle boundary defects, refinement of microstructure and of course, to get the required shape. However, the processing parameters are different from those of the respective matrix alloys and were identified by generating processing maps based on the dynamic materials model for a range of these MMCs[4-10]. The interface between the matrix and the reinforcement plays an important role in determining the final properties of the composite. The major findings of studies undertaken at DMRL and Northwestern University, USA were highlighted in a recent review paper[11]. The high temperature creep resistance of these MMCs was also the subject of an extensive study[12-18]. $MoSi_2$ matrix composite reinforced with SiC powder was produced by the reaction hot pressing route and characterized[19]. This paper outlines some of the significant results covering the various aspects relating to the dimensionally stable composites, processing maps, interface characteristics, high temperature creep behaviour and the $MoSi_2$ based composite.

II. DIMENSIONALLY STABLE COMPOSITES

Dimensional stability is one of the important properties that requires to be met in applications such as laser mirror substrate or inertial guidance control system components. The two main design parameters in this connection are the microyield strength and the micro creep resistance. The micro yield strength (MYS) is the stress required to cause a plastic strain of 10^{-6} m/m[20] and the micro creep refers to time dependent strains of small magnitude, occurring under conditions where creep is not considered likely, i.e., temperature near room temperature or below and stresses that are low relative to conventional yield strength values. Significant micro creep can occur at stresses well below MYS. MMCs have been found to have significantly higher micro yield strengths and better micro creep resistance as compared to beryllium which is the commonly used dimensionally stable material. Figure 1 shows the comparison of the microcreep behaviour of 2124 Al based MMCs with that of Be[21]. It also shows the importance of the mean SiC particle size and the volume fraction of reinforcement on the micro creep behaviour of the composite. The MMC containing a mean SiC particle size of <3.5 µm exhibits superior microcreep resistance not only with respect to Be but also to the MMC containing a coarser mean particle size of 5µm. For a given mean SiC particle size of <3.5 µm, the increase in the volume fraction of SiC from 10 to 30 % results in a significant improvement in the microcreep resistance of the composite. Table I gives the comparison of other important properties including the MYS.

The process route developed for PM processing of these composites involving deagglomeration of the reinforcement powder, blending the deagglomerated reinforcement powder with the matrix alloy powder, degassing the powder mixture, cold isostatic pressing and vacuum hot pressing has been optimized[2] and the process leads to a vacuum hot pressed billet with typical microstructure shown in Fig.2(a). The microstructure consists of a necklace like distribution of the reinforcement particulate around the prior matrix particle boundaries and this distribution can be further refined for improving the mechanical properties. Some amount of improvement in distribution can be achieved by changing the size ratios of the matrix and reinforcement particulate but substantial improvement is possible only by deformation processing of these composites. Fig.2(b) shows the typical microstructure after extrusion and the Table II summarizes the consequent increase in strength values.

III. PROCESSING MAPS

The processing parameters for these composites differ from those of conventional alloys and hence the forming temperatures and deformation rates have to be optimized for each composite system. In recent years, processing maps based on the dynamic materials model[22] are being adopted as a scientific tool for the optimization of hot working parameters as opposed to the trial and error methods conventionally employed for this purpose. Processing maps have been used to study the influence of initial microstructure, the matrix alloy and the reinforcement volume fraction on the hot working characteristics of DRA composites. Processing maps have also been employed for identifying the optimal processing parameters. Initially commercial purity aluminium was chosen as the matrix material to get a feel for the powder metallurgy processing of these composites and they were also used to study the effect of processing history. Volume fraction of reinforcement is an important parameter which influences the mechanical properties of these composites, especially, the strength, ductility and the fracture toughness. 2124 Al alloy was selected for further studies involving the effect of particulate addition as it finds important applications in defence and aerospace fields due to its superior combination of fracture toughness and strength when compared to existing ingot metallurgy and other PM alloys. 2014 Al and 6061 Al alloy based composites produced by the liquid metallurgy route by DURALCAN, USA, were also investigated for the purpose of comparison and to see the effect of matrix alloy on the processing maps.

Processing maps have been generated for 1100 Al, 2124 Al, 2014 Al and 6061 Al based composites having SiC or Al_2O_3 particulate reinforcement. Fig.3 shows a typical processing map for 2124 Al-20 vol. % SiC_p composite in the hot pressed condition. The different domains of dominating deformation mechanisms have been identified and the resulting microstructures shown. The specimens in the as-sintered, as-hot pressed and as-extruded conditions were evaluated for their hot working characteristics. Different volume fractions of reinforcements varying from 0-20 % were studied. The processing maps generated were used for both identifying the optimal process parameters as well as to understand the different deformation mechanisms operating under different processing conditions.

The processing maps for 1100 Al-10 vol. % SiC composite in both the sintered condition and the extruded condition revealed that the peak of the dynamic recrystallization (DRX) domain gets shifted from 0.01 s^{-1} in the sintered condition to 1.0 s^{-1} after extrusion[4]. Similar effect on the DRX domain was observed on reinforcement addition, in the processing maps for 2124 Al based composite and the matrix material[5]. The shifting of the DRX domain to higher strain rates on deformation processing and on reinforcement addition, has been explained in terms of the changes in the microstructure, especially the generation of the dislocation substructure and refinement in subgrain size. Dynamic recrystallization is controlled by the rate of migration of interfaces in the aluminium alloy matrix[23], which is enhanced by the fine dislocation substructures generated by the presence of hard particles and hence the DRX domain shifts to higher strain rates. The domain of dynamic recrystallization (DRX) is ideal for billet conditioning of powder metallurgy DRA composites as it helps eliminate the prior particle boundary (PPB) defects. The 2124 Al-20 vol. % SiC composite also exhibited a domain of superplasticity at 550°C and 0.001 s^{-1} [5]. The domain of superplasticity was found in 2014 Al-20 vol. % Al_2O_3 (DURALCAN) composite as well[6], while the map for 6061 Al-10 vol. % Al_2O_3 (DURALCAN) composite did not show a domain of superplasticity[7].

DRA composites generally exhibited microstructural instability at higher strain rates employed for processing conventional aluminium alloys. 2124 Al and 2014 Al based composites undergo superplastic deformation at orders of magnitude higher strain rates than conventionally superplastic aluminium alloys, while 6061 Al based composite did not show a domain of superplasticity. A criticality is observed at a reinforcement content of 10 volume percent[8]. The composite containing 10 volume per cent of reinforcement shows a tendency for abnormal grain growth and as a consequence the DRX domain shifts to lower strain rates and the superplasticity domain disappears. However, at higher strains the DRX process leads to finer grains and the DRX domain shifts back to higher strain rates. Superplasticity domain also appears back in the processing map.

Optimal processing parameters have been established for the composites studied, using the processing maps approach. Processing maps based on the dynamic materials model provide a scientific method to optimize the process parameters and is very useful in understanding the hot working characteristics of complex materials, such as advanced composites evaluated in this study.

IV. CREEP BEHAVIOUR

The reported work on the creep behaviour of aluminium matrix composites indicates that the understanding of the creep mechanisms is rather poor and a number of anomalies related to the existence of a proper steady state stage, operative creep mechanisms etc. exist. Therefore, a comprehensive study on the creep behaviour of particulate reinforced aluminium matrix composites was undertaken to resolve some of these issues[12-18]. In this work, four different aluminium matrix composites, viz., Al/SiC, Al-Mg/SiC, Al/TiB$_2$ and Al/TiC produced by powder metallurgy route were used. Al/TiB$_2$ was also produced by the XD route. Tensile and compressive creep data of Al-10 vol. % SiC (1.7µm) suggested that the steady state creep rate data of compression does not correspond to the minimum creep rate data of tension above a certain stress, referred to as "transition stress" because of the early onset of the tertiary stage in tensile creep. It is therefore concluded that only the compression creep rate data can be used for meaningful evaluation of the steady state creep mechanisms in composites[12,13].

To identify the operative creep mechanisms in composites, the matrix was chosen as pure aluminium to avoid complication due to solid solution and precipitation effect. Al/SiC composites with different particle sizes of SiC from 1.7 to 45.9 µm (keeping the volume fraction constant at 10 %) and different volume fractions from 10 to 30 % (keeping the constant particle size of 1.7 µm) were studied. The steady state creep rate data of all these composites obtained from compression exhibited very high values of stress exponent, $n \sim 20$ and an activation energy, $Q = 257$ kJ mol^{-1}, which do not conform to any theoretical or phenomenological models for dislocation creep. It was demonstrated that the present data for fine (1.7 µm) particulate composites can be rationalized with the substructure invariant model, which predicts a stress exponent of 8 and a subgrain size exponent of 3 and an activation energy close to the activation energy for lattice self diffusion together with the existence of a threshold stress. On the other hand, the creep rate data for coarse particulate composites (14.5 and 45.9 µm) agree well with the stress dependent substructure model with the stress exponent of 5 and lattice diffusion control[14].

A new form of dislocation creep mechanism map which plots normalized subgrain size as a function of inverse of normalized effective stress was constructed (Fig.4) to show the various pertinent dislocation creep mechanisms and the regimes of their dominance. As can be seen from this map if the interparticle spacing is more than the stress independent subgrain size then the stress dependent substructure mechanism is operative with a stress exponent of 5. The data of Al-10 vol. % SiC_p composite with 14.5 and 45.9 µm sizes were found to be consistent with this mechanism. Whereas, for the composites with finer particle size of 1.7 µm, constant substructure model is operative with stress exponent of $8^{[14]}$.

The effect of solid solution on the steady state creep of Al/SiC composite was studied by adding Mg as an alloying element. The summary of all the creep data is presented in Fig.5, which shows clearly the effect of reinforcement on improving the creep resistance. The creep rate of Al-4Mg-10 vol. % SiC is very close to that of Al-20 vol. % SiC, suggesting that the effect of solid solution strengthening by 4Mg is approximately equal to composite strengthening by 10 vol. % of SiC[15]. The role of reinforcement other than SiC on the steady state creep behaviour of composite was studied by choosing TiB_2 particulates[16,17]. This composite is attractive not only because of the higher modulus of TiB_2 (which is expected to give higher creep strength) but also because this composite can be produced by the XD route which results in very fine particles of TiB_2 with better interface with aluminium which in turn leads to improved creep resistance (Fig.5).

Another system that has been investigated is Al/TiC wherein the Al and TiC phases are thermodynamically stable at the processing temperature which is well above that of molten Al. However, at lower temperatures they cannot coexist and prolonged exposure to temperature (873 K) has been found to result in the appearance of Al_3Ti and Al_4C_3 phases [24,25]. Substantial increases in Young's modulus and strength properties from the solid state reactions were observed, but there was a decrease in ductility. A component made by forming highly ductile XD Al/TiC composite may be heat-treated to give much higher stiffness, hardness and strength. Fig.6 shows the free energy versus temperature curves for reactions:

$$3TiC + 13\ Al(l/s) = Al_4C_3 + 3Al_3Ti \tag{1}$$

$$3TiC + 7\ Al(l/s) = Al_4C_3 + 3TiAl \tag{2}$$

$$3TiC + 4\ Al(l/s) = Al_4C_3 + 3Ti \tag{3}$$

Reaction (1) proceeds to the right only below 1025 K, above which the free energy change for the reaction is positive. The presence of Al_3Ti and Al_4C_3 after heat treating at 913 K is expected. Reactions (2) and (3) have positive free energy changes at all temperatures. The evolution of the microstructure can be discussed further using the ternary Al-Ti-C phase diagram in Fig.7 drawn for 913K from the binary phase diagrams of the Al-C, Ti-C and Al-Ti systems and the ternary phase diagram at 1023K which is just above T_m of Al. The ternary compounds $H(Ti_2AlC)$ and $P(Ti_3AlC)$ were assumed not to change on cooling to 913K. Tie lines between coexisting phases have been drawn using information from Fig.6. TiAl and Al_4C_3 are not joined, as equation (2) shows that they react to form Al and TiC at 913K. Also, the chemical reaction:

$$13TiAl + 2Al_4C_3 = 6TiC + Al_3Ti \tag{4}$$

has a negative free energy change up to temperatures higher than 5000K, predicting that TiC and Al_3Ti coexist. Similarly, Ti_3Al and Al_4C_3 are not joined. The present calculations agree well with the corrected partial phase diagram of Norman et al[26] and the predictions from the three dimensional chemical potential diagrams of Al-Ti-C system drawn by Yokokawa et al for 973K. The composition in the present case lies in triangle I in Fig.7, where Al, Al_3Ti and Al_4C_3 are the stable phases. The effect of heat treatment on the mechanical properties of the Al-20 vol. % TiC is shown in Table III and that on creep rate of the composite is shown in Fig.8[18].

V. INTERFACES

The strength, stiffness and fracture characteristics of MMCs are critically dependent on the interface between the matrix and the interface. A 'good' interface which implies a strong chemical bond across the interface, efficiently transfers the stresses from the ductile matrix to the stronger and stiffer reinforcement. The elastic modulus of the composite is an indirect measure of the strength of the matrix/reinforcement interface bond. Fig.9 shows the ratio of the experimental value of the elastic modulus and the elastic modulus predicted by the rule of mixture (ROM) model[11] for Al matrix composites reinforced with 20 vol. % of particulates of different ceramics namely SiC, B_4C, TiC and TiB_2. A well bonded interface would allow the maximum transfer of load from the matrix to the reinforcement and is expected to result in a high value of E(Expt.)/E(ROM) ratio. From Fig.9 it may be inferred that among the reinforcements mentioned the Al/TiC bond is the strongest and Al/TiB_2 is the weakest. Interface bonding may be classified as mechanical or chemical. Mechanical bonding is significant only in the case of fibre reinforced composites, when fibres have rough or faceted surfaces. Chemical bonding is important for all kinds of reinforcements, viz. fibres, whiskers and particulates. A chemical bond is possible only if the atoms of the matrix and the reinforcements are in direct contact and is accomplished by an exchange of electrons and the type of exchange determines the character of the bond. It can be metallic, which is non-directional, and ionic or covalent which are directional. An interface with a metallic bond is thus more ductile than that with ionic or covalent bonds. Fig.10 shows for four MMC systems with reinforcements having different types of bonds, a bar diagram of E(Expt.)/E(ROM)[11]. It is seen that metal-intermetallic and metal-metal systems show higher values of E(Expt.)/E(ROM) as compared to metal-ceramic systems. Also Al/TiC bond is stronger compared to Al/SiC, Al/TiB$_2$ and Al/B$_4$C. In metal-ceramic bonding, transition metal carbides like TiC, which have partial metallic nature thus show stronger bond with Al, a reactive metal than more covalent bonded ceramics like TiB$_2$ (Table IV). It is also clear from Table IV that TiB$_2$ has a higher heat of formation and is more stable. Also TiC reacts with Al under certain conditions while TiB$_2$ does not. Thus, higher stability of TiB$_2$ in Al as compared to that of TiC is another reason, why the latter is wetted better.

VI. MoSi$_2$/SiC COMPOSITES

Silicide matrix composites possess a strong potential for high temperature structural applications for temperatures above 1200°C. MoSi$_2$ is quite brittle like ceramics at room temperature, but shows significant deformation at temperatures of 1000-1200°C and above. It has outstanding oxidation resistance up to a temperature as high as 1700°C, except at a temperature range of 400-600°C it has been found to oxidize. Porous MoSi$_2$ disintegrates completely on exposure to air at 500°C and the phenomenon is called pesting. The method adopted to improve room temperature fracture toughness and the yield strength and creep resistance at high temperature is to reinforce the matrix with ceramic particles or whiskers.

Reaction hot pressing process (Fig.11) has been developed at DMRL to obtain MoSi$_2$ from an intimate mixture of Mo and Si elemental powders. In the case of composite, measured quantity of SiC powder was blended with Mo-Si mixture[19]. Hot pressing was done at 1450°C for two principal reasons: (i) availability of liquid Si phase (M.P.=1410°C), which enhances the reaction rate with Mo and densification and (ii) favourable thermodynamic parameters[27] for complete reaction. The density measurements showed that more than 98% of theoretical density was achieved. The results of XRD and EPMA have proved that reaction during hot pressing is complete and only $MoSi_2$ forms. The SEM back scattered electron micrograph of $MoSi_2$/20 SiC is given in Fig. 12(a). The coarse size (30 μm) of the Mo particles is responsible for the nonuniform distribution of SiC particles. Fig. 12(b) shows an optical micrograph recorded using polarised illumination revealing $MoSi_2$ grains. Fig. 13(a) and (b) are bright and dark field TEM micrographs of two different $MoSi_2$ grain boundary/ SiC triple point junctions, showing absence of porosities or amorphous SiO_2 phase. High resolution transmission electron microscopy has further established that there is no amorphous silica film at the $MOSi_2$ grain boundaries. Fig. 14 shows atomic resolution TEM images: (a) $MoSi_2$ grain boundary, which is parallel to close packed and low index $MoSi_2$ (110) plane; (b) $MoSi_2$/SiC interface, where atoms of both the phases are in direct contact[19]. The interface is faceted on an atomic scale and facets are parallel to SiC densely packed (200) planes. Thus bonding is expected to be strong. However, not all interfaces are atomically abrupt. Some show an amorphous film 5-8 nm thick. This is shown in Fig. 15[19]. The composition of this amorphous phase is yet to be determined. The mechanical properties of the composite as well as that of the polycrystalline $MoSi_2$ obtained from literature for the purpose of comparison are shown in Table V. Further improvement in room temperature bend strength and fracture toughness can be obtained with more uniform distribution of the SiC particles. Variation of hardness with increase in temperature is plotted in Fig. 16[19]. The hardness data for the composite has been compared with hardness data for $MoSi_2$ and that with 2 wt.% carbon, prepared by hot pressing $MoSi_2$ powders[28], mechanically alloyed and HIPed $MoSi_2$[29]. The composite is harder than as hot pressed $MoSi_2$ at all temperatures. The mechanically alloyed and HIPed $MoSi_2$ is harder at temperatures up to 1000°C, possibly due to finer grain size. At temperatures higher than 1000°C, the composite is harder. The indented microstructures were observed at different temperatures. Indentation cracks could be noticed up to a temperature of 1000°C, but not at 1100°C and above, indicating the existence of ductile to brittle transition. The compressive yield strength data as a function of temperature for the composite is compared with that of the matrix material obtained from literature[30] in Fig.17.

VII. SUMMARY

The powder metallurgy processing of discontinuously reinforced aluminium alloy MMCs and $MoSi_2$ matrix composites have been developed at DMRL and the composites have been evaluated for various properties. Processing maps have been generated for a variety of composites and the hot working parameters optimized. Creep properties have been evaluated and a dislocation creep mechanism map proposed for particle-reinforced materials. Interface studies carried out on different matrix/reinforcement systems have been analyzed. Composites have been developed for applications requiring dimensional stability as one of the main design parameter.

ACKNOWLEDGEMENTS

The authors are grateful to Director, D.M.R.L., for his encouragement and permission to publish this work. The authors would like to acknowledge the kind cooperation recieved from Dr.A.K.Gupta, National Physical Laboratory, New Delhi; Dr. C. Ganguly, Bhabha Atomic Research Centre, Bombay and Professor Y.V.R.K.Prasad, Indian Institute of Science, Bangalore. One of the authors (R.M.) is grateful to Dr. Wen-An Chiou, Associate Research Professor, Department of Materials Science and Engineering, Northwestern University, Evanston, Illinois, USA for high resolution TEM work. The assistance obtained from other members and technical staff of the Composites Activity Team of DMRL is also gratefully acknowledged.

REFERENCES

1. Y.R. Mahajan, V.V. Bhanuprasad, B.V.R. Bhat and A.B. Pandey: *Principles of Solidification and Materials Processing*, vol.2, Oxford & IBH Pub. Co. Pvt. Ltd., New Delhi, India, 1989, pp. 613-630.

2. V.V. Bhanuprasad, B.V.R. Bhat, A.B. Pandey, K.S. Prasad, A.K. Kuruvilla and Y.R. Mahajan: *The Int. Jr. Pow. Met.*, 1991, vol. 27(3), pp. 227-235.

3. Manoj K. Jain, Velidandla V. Bhanuprasad, Samir V. Kamat, Avadh B. Pandey, Vijay K. Varma, B.V. Radhakrishna Bhat and Yashvant R. Mahajan: *The Int. Jr Pow. Met.*, 1993, vol. 29(3), pp. 267-275.

4. B.V. Radhakrishna Bhat, Y.R. Mahajan, H.Md. Roshan and Y.V.R.K. Prasad: *J. Mater. Sci.*, 1993, vol. 27, pp. 2141-2147.

5. B.V. Radhakrishna Bhat, Y.R. Mahajan, H.Md. Roshan and Y.V.R.K. Prasad: *Metall. Trans. A*, 1992, vol. 23A, pp. 2223-2230.

6. B.V. Radhakrishna Bhat, Y.R. Mahajan, H.Md. Roshan and Y.V.R.K. Prasad: *Mater. Sci. Engg. A*, 1994, vol. 189, pp. 137-145.

7. B.V. Radhakrishna Bhat, Y.R. Mahajan, H.Md. Roshan and Y.V.R.K. Prasad: *Accepted for publication in Mater. Sci. Technol.*, 1994.

8. B.V. Radhakrishna Bhat: *Ph.D. Thesis*, 1994, Metallurgical Engineering Dept., Indian Institute of Technology, Madras, India.

9. B.V. Radhakrishna Bhat, Y.R. Mahajan and Y.V.R.K. Prasad: *Submitted to Acta Metall. Mater.*, 1994.

10. B.V. Radhakrishna Bhat, Y.V.R.K. Prasad and Y.R. Mahajan: *Accepted for publication in the special volume of Key Engineering Materials series on MMCs*, Trans Tech Pub. Ltd., Switzerland, 1995.

11. Rahul Mitra and Yashwant R. Mahajan: *Defence Science Journal*, vol. 43(4),

1993, pp. 397-418.

12. A.B. Pandey, R.S. Mishra and Y.R. Mahajan: *Scripta Metall. Mater.*, 1990, vol. 24, pp. 1565-1570.

13. A.B. Pandey, R.S. Mishra and Y.R. Mahajan: *J.Mater.Sci.*, 1993, vol. 28, pp.2943-2949.

14. A.B. Pandey, R.S. Mishra and Y.R. Mahajan: *Acta Metall.*, 1992, vol. 40, pp.2045-2052.

15. A.B. Pandey, R.S. Mishra and Y.R. Mahajan: *Submitted to Metall. Trans. A.*, 1995.

16. A.B. Pandey, R.S. Mishra and Y.R. Mahajan: *Mater. Sci. Eng. A*, 1994, vol. 189, pp. 95-104.

17. A.B. Pandey, R.S. Mishra and Y.R. Mahajan: *Scripta Metall. Mater.*, 1993, vol.29, pp. 1199-1204.

18. A.B. Pandey, R.S. Mishra and Y.R. Mahajan: *Submitted to Mater. Sci. Eng.A*, 1995.

19. R. Mitra, Y.R. Mahajan, N.E. Prasad, W.A. Chiou and C. Ganguly: *Accepted for publication in the special volume of Key Engineering Materials series on MMCs*, Trans Tech Pub. Ltd., Switzerland, 1995.

20. Walter R. Mohn and Gerald A. Gegel: *Advanced Composites-The Latest Developments*, Proceedings of the second conference on Advanced Composites, Dearborn, Michigan, ASM International, 1986, pp.69-73.

21. W.R. Mohn and D. Vukobratovich: *SAMPE J.*, vol.24(1), 1988, pp.26-34.

22. Y.V.R.K. Prasad, H.L. Gegel, S.M. Doraivelu, J.C. Malas, J.T. Morgan, K.A. Lark and D.R. Barker: *Metall. Trans.A*, vol. 15A, 1984, pp. 1883-1892.

23. Y.V.R.K. Prasad and N. Ravichandran: *Bull. Mater. Sci.*, vol. 14, 1991, pp. 1241.

24. R. Mitra, M.E. Fine and J.R. Weertman: *J. Mater. Res.*, vol. 8(9), 1993,

pp.2370-2379.

25. K. Satyaprasad, Y.R. Mahajan and V.V. Bhanuprasad: *Scripta Metall.*, vol.26, 1992, pp.711-716.

26. J.H. Norman, G.H. Reynolds and L. Brewer: *Intermetallic Matrix Composites*, Mater. Res. Soc. Symp. Proc., vol. 194, Pittsburgh, PA, 1990, p.369.

27. I. Barin: *Thermochemical Data of Pure Substances*, vol.2, Weinheim, Germany: VCH Verlagsgesellschaft mbH, D-6940, 1989, p.931.

28. S.A. Maloy, J.J. Lewandowski, A.H. Heuer and J.J. Petrovic: *Mater. Sci. Eng. A*, vol. 155, 1992, p.159.

29. S.N. Patankar and J.J. Lewandowski: *High Temperature Ordered Intermetallic Alloys V*, Mater. Res. Soc. Symp. Proc., vol. 288, Pittsburgh, PA, 1993, pp.829-834.

30. R.M. Aikin, Jr.: *Ceram. Eng. Sci. Proc.*, vol. 12(9-10), 1991, pp. 1643-1655.

Table I. Properties of dimensionally stable MMCs and conventional materials.

PROPERTIES	Be	2124Al (T6)	6061Al (T6)	2124Al+ 30v/oSiC	6061Al+ 35v/oSiC
density (g/cc)	1.85	2.78	2.70	2.91	2.88
E (GPa)	303	72	69	117	131
MYS (MPa)	35	117	110-131	117	124
CTE (x 10^{-6}/K)	11.5	23.4	23.0	12.4	12.1
Thermal Conductivity (W/mK)	201	154	171	122	--

Table II. Typical mechanical properties of 2124 Al-30 vol. % SiC_p composite.

COMPOSITION	PROCESSING CONDITION	E (GPa)	U.T.S (MPa)
2124 Al-30 vol. % SiC_p	Vacuum Hot Pressed(VHP)	108	435
2124 Al-30 vol. % SiC_p	VHP+Extruded	120	497

Table III.   Effect of heat treatment on the mechanical properties of Al-20 vol. % TiC composite.

CONDITION	E(GPa)	Y.S.(MPa)	U.T.S.(MPa)	%ELONG.	VHN
HOT ROLLED	105	173	239	6.84	73
600°C/96Hrs	129	308	348	0.8	156

Table IV.   A comparison of bonding, reactivity with Al and wetting characteristics of TiC and TiB$_2$.

COMPOUND	ΔH_f (kJ/mol)	Covalent bond (%)	Reactivity with Al ΔG_f (kJ)	Reaction Product	Contact Angle (Cosθ) with Al
TiC	10.5	80	-5	Al$_3$Ti, Al$_4$C$_3$	0.7
TiB$_2$	19.6	92	44.6	Al$_3$Ti, AlB$_2$	0.4

Table V. Comparison of mechanical properties of $MoSi_2$-20 vol. % SiC composite and polycrystalline $MoSi_2$.

PROPERTY	$MoSi_2$-20 vol.% SiC_p	Polycryst. $MoSi_2$
Vicker's Hardness (GPa)	11.3	9.8
3 Point Bend Strength (MPa)	269	140-160
Fracture Toughness Indentation(MPa$\sqrt{m}$)	7.5	3.0-4.0

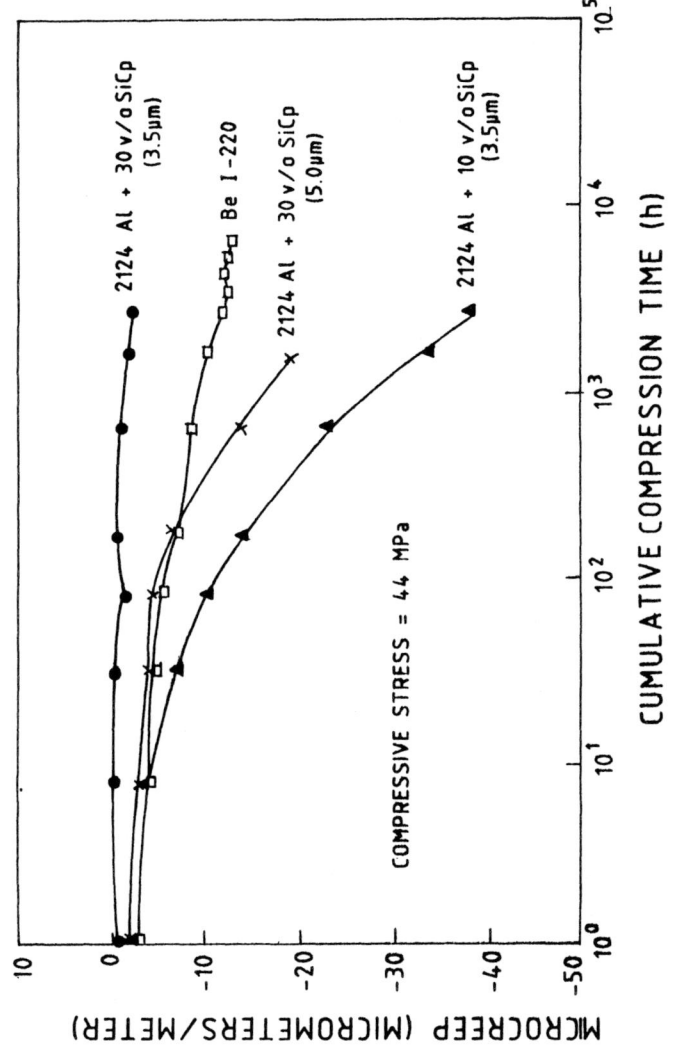

Fig.1.Comparison of microcreep behaviour of MMCs and Be[21].

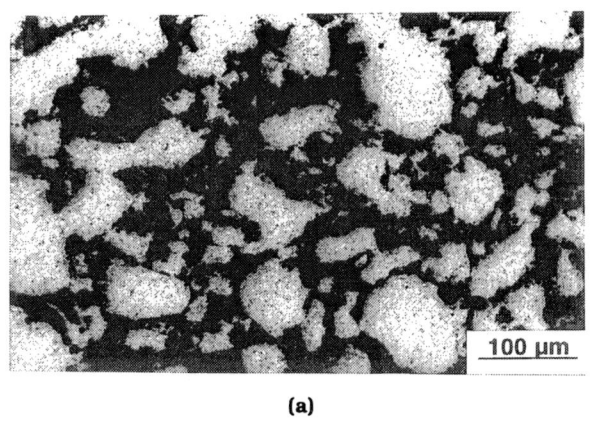

(a)

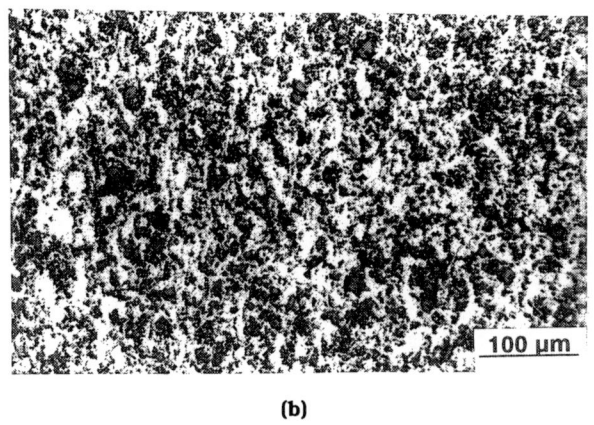

(b)

Fig.2. Typical microstructures of 2124 Al/SiC composite[3] (a) as-hot pressed condition (b) as-extruded condition.

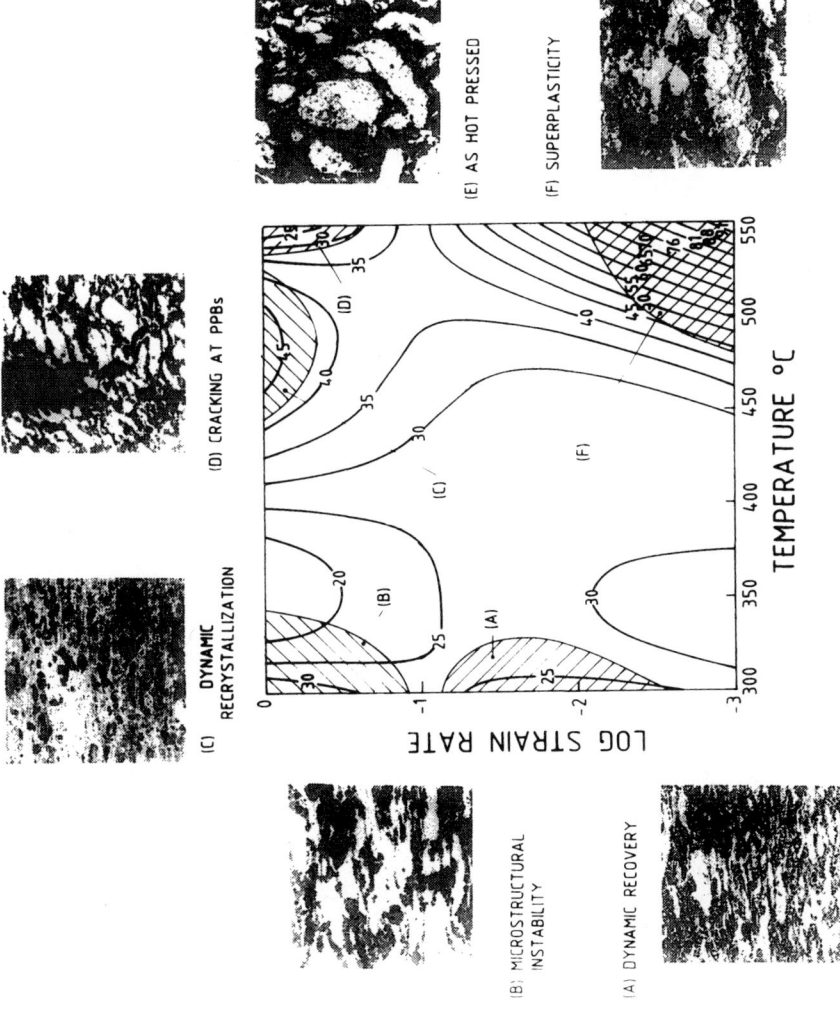

Fig.3. Typical processing map for 2124 Al-20 vol. % SiC$_p$ in the as-hot pressed condition[5].

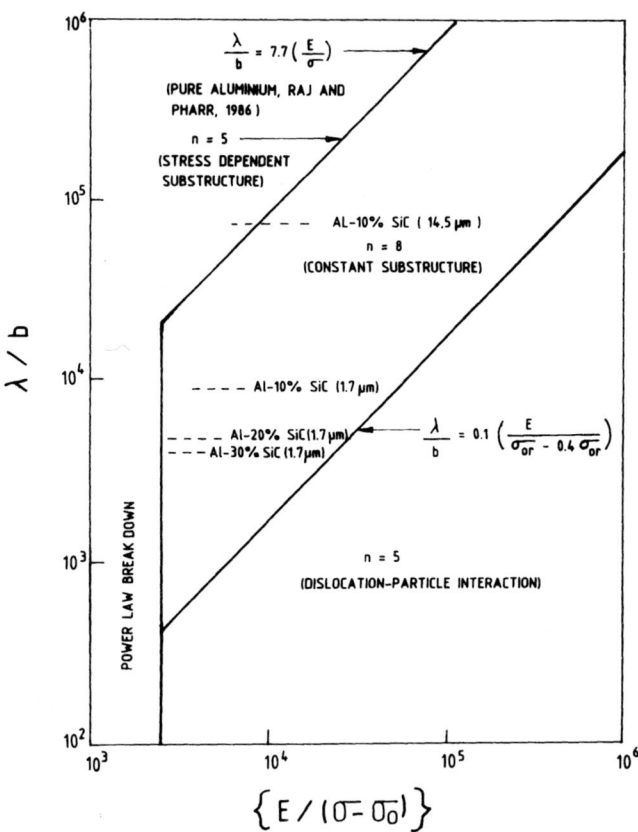

Fig.4. Dislocation creep mechanism map for the particle-reinforced materials[14].

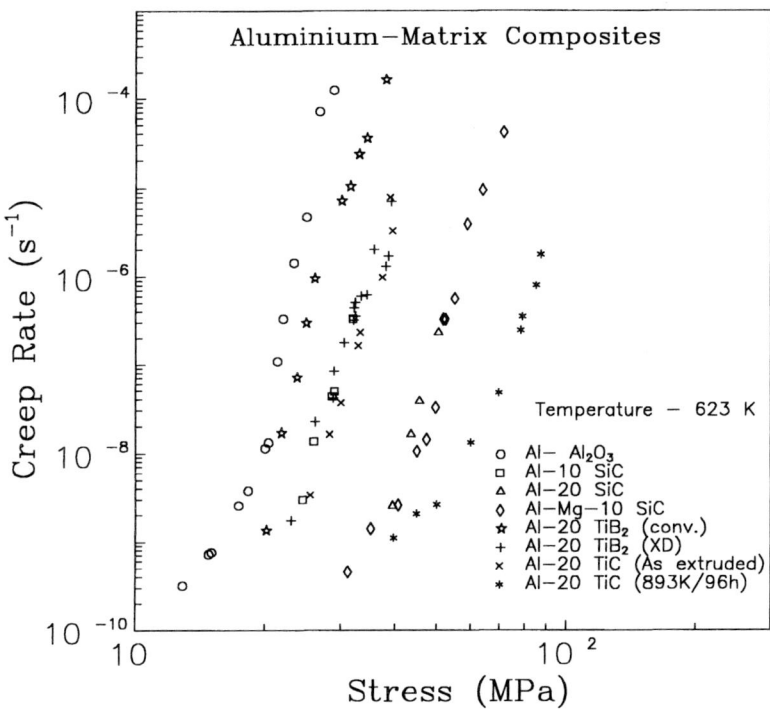

Fig.5. Summary of steady state creep data for all the MMCs and the matrix material studied.

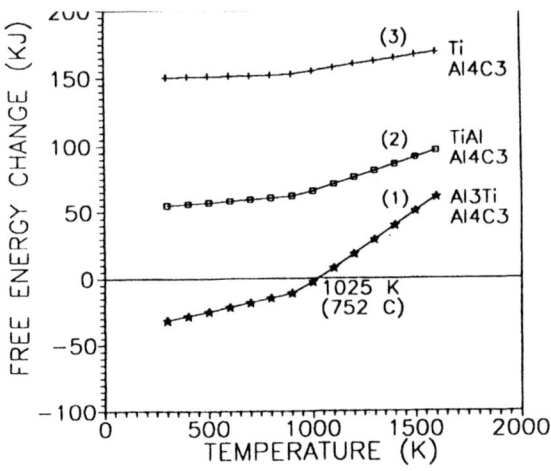

Fig.6. Free energy versus temperature plots for reactions(1)-(3)[24].

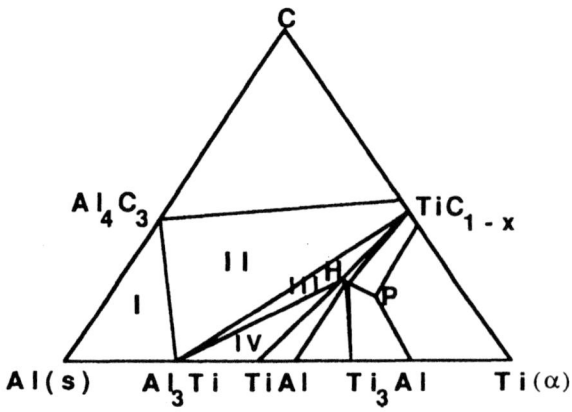

Fig.7. Isothermal ternary Al-Ti-C phase diagram at 913K[24].

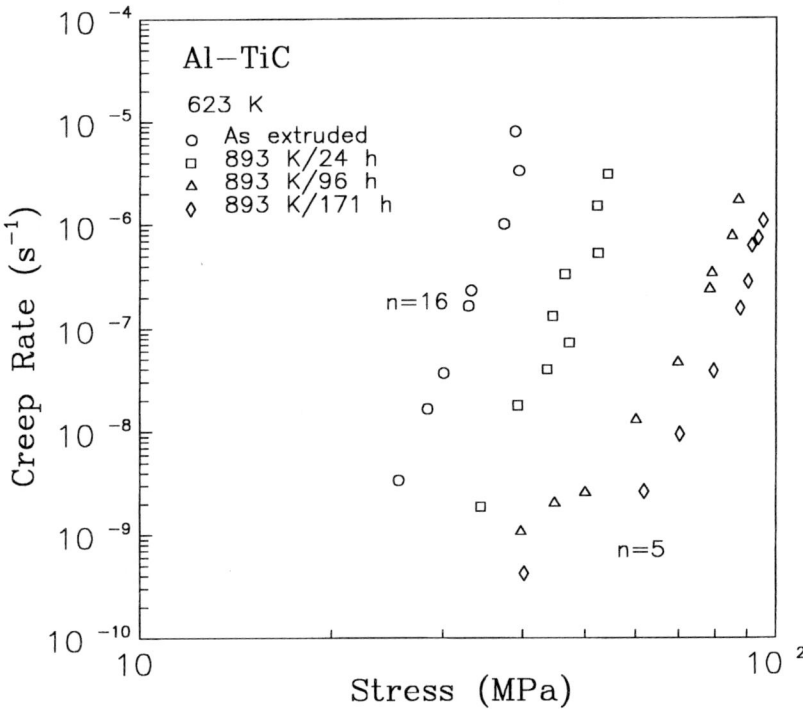

Fig.8. Steady state creep data for Al/TiC composite before and after isothermal heat treatment[18].

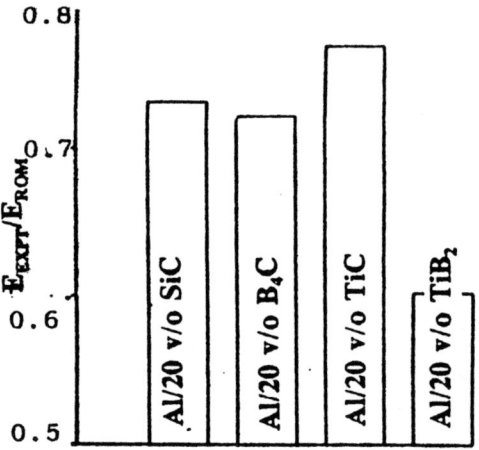

Fig.9. E_{EXPT}/E_{ROM} for Al matrix composites with different reinforcements[11].

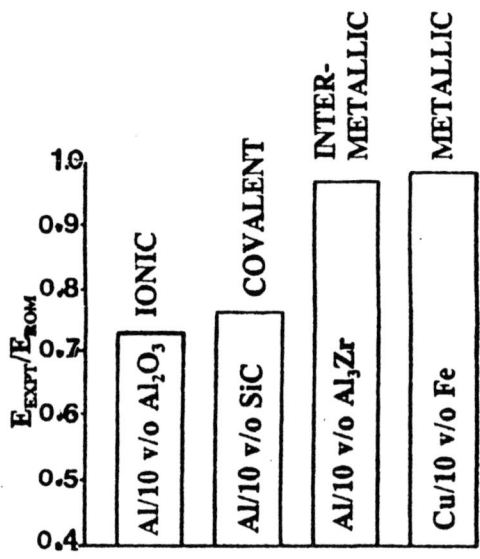

Fig.10. E_{EXPT}/E_{ROM} for four different MMC systems with reinforcements having different chemical nature[11].

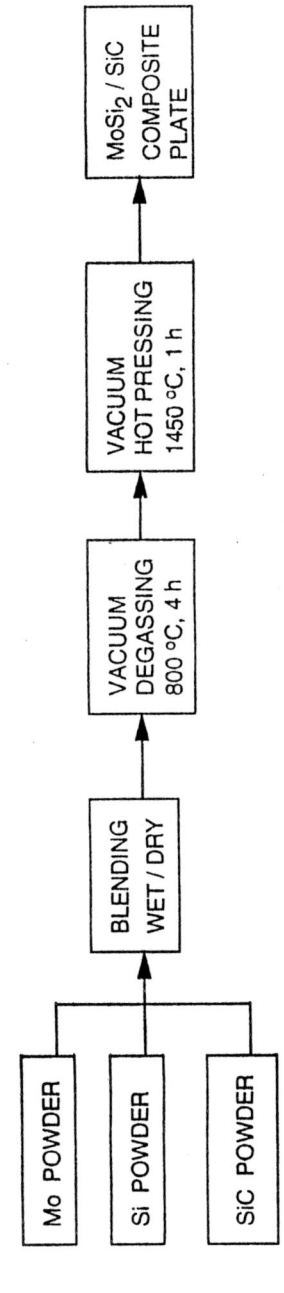

Fig.11.Reaction hot pressing process flow chart for the synthesis of $MoSi_2$/SiC composite[119].

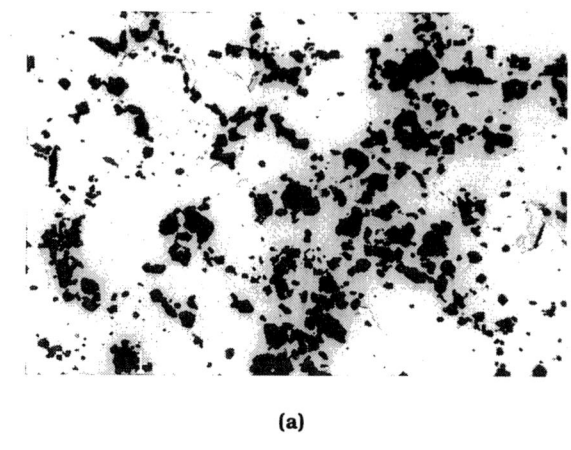

(a)

(b)

Fig.12. Microstructure of MoSi$_2$/SiC composite: (a) Backscattered electron image; (b)Optical micrograph using polarized illumination[19].

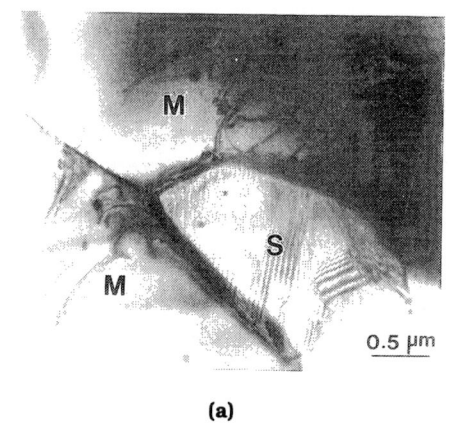

(a)

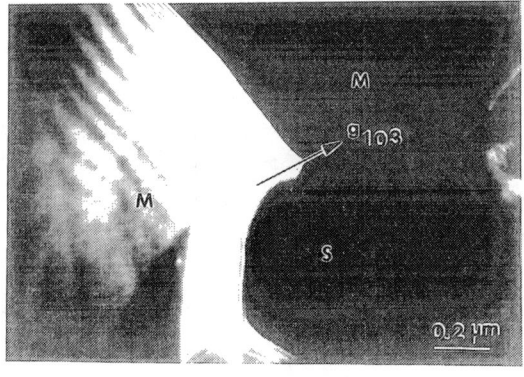

(b)

Fig.13. TEM micrographs of a MoSi$_2$ ('M') grain boundary/SiC$_p$ ('S') triple point junction, showing absence of porosities or amorphous SiO$_2$ phase: (a) Bright field; (b) Dark field.

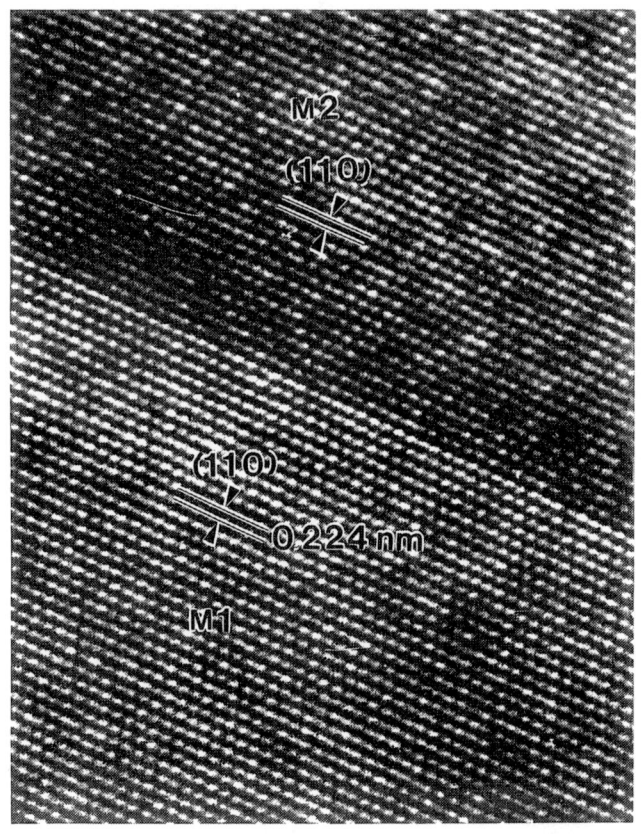

(a)

Fig. 14. Atomic resolution TEM images: (a) $MoSi_2$ grain boundary between grains M_1 and M_2; (b) $MoSi_2$/SiC interface. Arrows indicate position of monoatomic steps. Amorphous phase is missing[19].

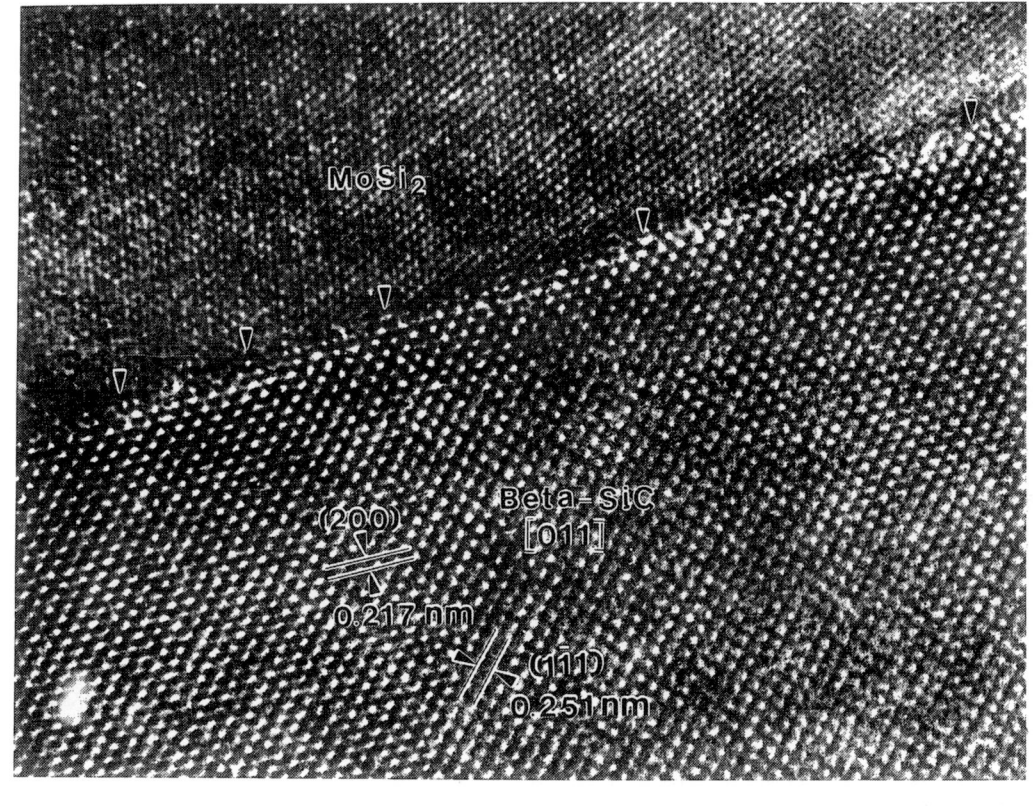

(b)

Fig. 14. Atomic resolution TEM images: (a) $MoSi_2$ grain boundary between grains M_1 and M_2; (b) $MoSi_2$/SiC interface. Arrows indicate position of monoatomic steps. Amorphous phase is missing[19].

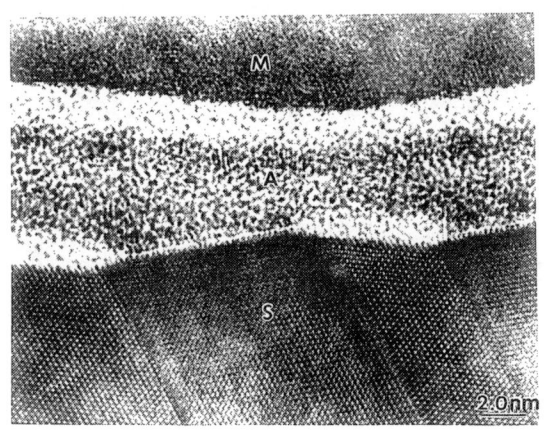

Fig.15. MoSi$_2$/SiC interface showing an amorphous phase, 5-8 nm thick[19].

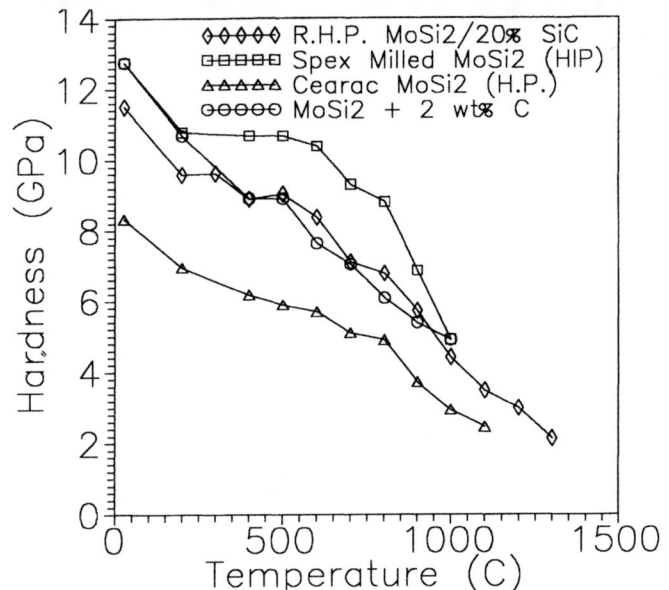

Fig. 16. Plots showing variation of hardness with temperature for monolithic MoSi$_2$[28,29] and reaction hot pressed MoSi$_2$/SiC composite[19].

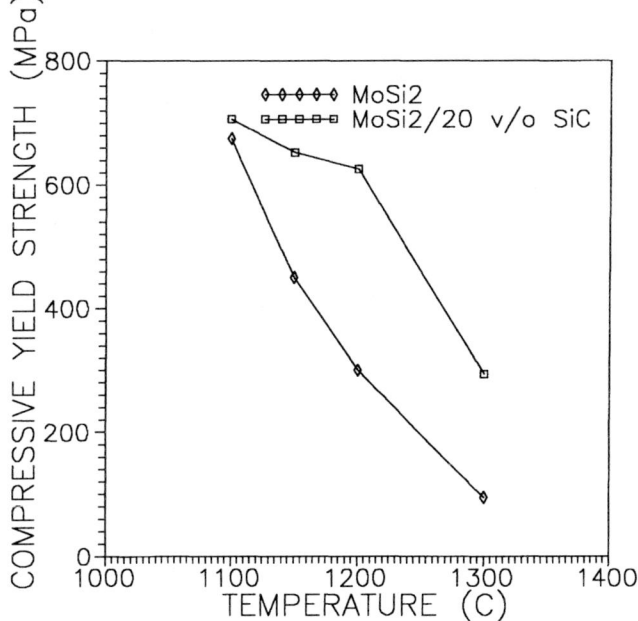

Fig. 17. Comparative plot of compressive yield strength as a function of temperature for MoSi$_2$/SiC composite and monolithic MoSi$_2$[30].

AUTHORS

Anandani, R. C., 245

Bhanuprasad, V. V., 269
Bhat, B. V. R., 269

Choh, T., 31

Dhar, A., 245
Dutta, B., 15

Ee, A. S., 109
El-Mahallawy, N. A., 91

Gupta, A. K., 245
Gupta, M., 109

Hunt, W. H., 155

Inem, B., 41

Jain, M. K., 269
Jayaram, V., 193

Kumar, S., 193

Lai, M. O., 109
Lee, W. B., 165
Lesuer, D. R., 155
Lewandowski, J. J., 155
Lu, L., 109

Mahajan, Y. R., 269
Malik, I. A., 245
Mani, T. V., 193
Mannikar, S., 15
Mitra, R., 269

Osman, T. M., 155

Pai, B. C., 131
Pan, J., 59
Pandey, A. B., 269
Pillai, R. M., 131
Pramila Bai, B. N., 171

Ranganath, S., 227
Ray, S., 69

Saifullah, M. S. M., 193
Saravanan, R. A., 171
Sarkar, J., 193
Satyanarayana, K. G., 131
Sikand, R., 245
Subrahmanyam, J., 207
Suery, M., 1
Sugamata, M., 121
Surappa, M. K., 15, 171

Taha, M. A., 91

Wan, H., 59
Warrier, K. G. K., 193

Yang, D. M., 59
Yao, Y. P., 165